TK
7867
.H462
2004

Schematic Capture with Electronics Workbench Multisim®

Marc E. Herniter

Associate Professor
ECE Department
Rose-Hulman Institute of Technology

Upper Saddle River, New Jersey
Columbus, Ohio

Editor in Chief: Stephen Helba
Editor: Dennis Williams
Development Editor: Kate Linsner
Production Editor: Rex Davidson
Design Coordinator: Diane Ernsberger
Cover Designer: Marc Shumaker
Cover art: SuperStock
Production Manager: Pat Tonneman
Marketing Manager: Ben Leonard

This book was printed and bound by Courier Kendallville, Inc. The cover was printed by Phoenix Color Corp.

Copyright © 2004, 2000, 1997 by Pearson Education, Inc., Upper Saddle River, New Jersey 07458. Pearson Prentice Hall. All rights reserved. Printed in the United States of America. This publication is protected by Copyright and permission should be obtained from the publisher prior to any prohibited reproduction, storage in a retrieval system, or transmission in any form or by any means, electronic, mechanical, photocopying, recording, or likewise. For information regarding permission(s), write to: Rights and Permissions Department.

Pearson Prentice Hall™ is a trademark of Pearson Education, Inc.
Pearson® is a registered trademark of Pearson plc
Prentice Hall® is a registered trademark of Pearson Education, Inc.

Pearson Education Ltd.
Pearson Education Singapore Pte. Ltd.
Pearson Education Canada, Ltd.
Pearson Education—Japan

Pearson Education Australia Pty. Limited
Pearson Education North Asia Ltd.
Pearson Educación de Mexico, S.A. de C.V.
Pearson Education Malaysia Pte. Ltd.

10 9 8 7 6 5 4 3 2 1

ISBN: 0-13-049614-6

Warning and Disclaimer

This book is designed to provide tutorial information about Electronics Workbench Multisim. Every effort has been made to make this book as complete and accurate as possible, but no warranty or fitness is implied.

The information is provided on an "as is" basis. The author and Prentice Hall shall have neither liability nor responsibility to any person or entity with respect to any loss or damages arising from the information contained in this book or from the use of the disks and programs that may accompany it.

Trademarks

"Adobe" and "Acrobat" are registered trademarks of Adobe Systems Incorporated.

"Multisim" and "Electronics Workbench" are registered trademarks of Interactive Image Technologies Limited.

"Microsoft," "Win32s," "MS-DOS," and "Windows" are registered trademarks of Microsoft Corporation.

"IBM" is a registered trademark of International Business Machines Corporation.

"Sun" is a registered trademark of Sun Microsystems Incorporated.

"Open Windows" is a trademark of Sun Microsystems Incorporated.

To my wife, Corena,

my daughters, Katarina Alexis Sierra, Laina Calysta Dine'h, Darrian Iliana Francheska,

my parents, Mom and Dad,

my cats, Zak, Chester, Tipper, Kennedy, Blaise, and Beethoven

my dogs, Samantha, Sage, and Wolfee,

my larger fish, Sedona, Flounder, Jewel, and the others my daughters have yet to name.

Preface

This manual is designed to show students how to use the Multisim circuit simulation program from Electronics Workbench. It is a collection of examples that show students how to create a circuit, how to run the different analyses, and how to obtain the results from those analyses. This manual does not attempt to teach students circuit theory or electronics; that task is left for the main text. Instead, the manual takes the approach of showing students how to simulate many circuits found throughout the curriculum. An example could be the DC circuit shown below.

It is assumed that the student has been given enough information to analyze the circuit. This manual assumes that the student wishes to check his or her answers (or intuition) with this program. The student would construct the circuit as shown in Part 1 and then run either some of the DC simulations in section 3.A or the DC Sweep in section 4.A.

This manual was designed to be used by students for their entire educational career and beyond. Since the parts are arranged by analysis type, they contain a range of examples from circuits covered in first-semester circuit theory courses to senior-level amplifier and switching circuits. Sections that are too advanced for beginning students may be skipped without loss of continuity. All parts contain both simple circuits and advanced circuits to illustrate the analysis types. Sections do not have to be covered sequentially. Individual examples can be identified that apply to specific courses. However, the following sequence is suggested for first-time users. All beginning users should follow Parts 1 and 2 completely to learn how to draw and save schematics, and how to use the Postprocessor and Grapher. Multisim instruments such as the Multimeter and Oscilloscope are covered in the sections in which they are used. All students should follow some of the examples in Parts 3 and 4 that are relevant to the course and also cover a few of the examples that may apply to earlier courses (if any). The early examples in these parts have the most step-by-step detail of how to use the software. Part 6 covers time domain analysis. The first section discusses the use of the oscilloscope, so this section should be assigned before any of the following sections are covered.

This manual contains examples that apply to courses throughout the curriculum. Introductory circuits classes usually cover DC circuits, AC circuits with phasors, and transient circuits with a single capacitor (or inductor) and a switch. Examples are given to cover these types of problems. After reviewing the examples in this manual, a student should be able to simulate similar problems. A typical first electronics course may cover transistor biasing, amplifier gain, and amplifier frequency response. Examples of these analyses are also given.

Exercises are given at the end of each section. These exercises specify a circuit and give the simulation results. The students are encouraged to work these problems to see if they can obtain the same simulation results. The exercises are intended to give students practice in using the software, not to teach them circuits. My philosophy is that simulation software should be used only to verify one's own calculations or intuition. In my classes I assign problems that are worked by hand calculation, simulated with Multisim, and then tested in the lab. The students then compare the measured results to the hand calculations and Multisim simulations. Without hand calculation, it is impossible to know if the Multisim simulations are correct.

Software Included with the Manual

The CD-ROM contains all of the circuit files used as examples in this manual. If you have a problem with one of the circuits you are simulating by following the text, you can view the circuit on the CD-ROM and see how it differs from your circuit.

Comments and Suggestions

The author would appreciate any comments or suggestions on this manual. Comments and suggestions from students are especially welcome. Please feel free to contact the author using any of the methods listed below:
- **E-mail:** Marc.Herniter@IEEE.ORG
- **Phone:** (812) 877-8512
- **Fax:** (253) 369-9536
- **Mail:** Rose-Hulman Institute of Technology, CM123, 5500 Wabash Avenue, Terre Haute, IN 47803-3999

Acknowledgments

I would like to thank my students at Rose-Hulman Institute of Technology for giving me continued inspiration to improve this manual. Without their constant curiosity, this book would not be necessary. I would like to thank Scott Duncan and Joseph Koenig of Electronics Workbench and Dennis Williams of Prentice Hall for getting together on this project and making the book possible. I would also like to thank Luis Alves and Tien Pham of Electronics Workbench for answering my barrage of questions. Finally, I would like to express my deepest appreciation to my wife and kids, who no longer let me sit in front of my computer twenty-four hours a day.

Before You Begin

General Conventions

- This manual assumes that you have a two- or three-button mouse. The words *LEFT* and *RIGHT* refer to the left and right mouse buttons.
- All text highlighted in bold refers to menu selections. Examples would be **File** and **Run.**
- All text in capital letters refers to keyboard selections. For example, press the **ENTER** key.
- **All text in this font refers to text you will see on the computer screen. This applies to all text except menu selections**.
- `All text in this font refers to text you will type into the program.`
- The word "select" means "click the left mouse button on."

Keyboard Conventions

Throughout the manual many keyboard sequences are given as shortcuts for making menu selections. The explanations of these sequences will be given later. It is important to know the conventions used to specify the sequences.

- Many control key sequences will be specified. For example, **CTRL-R** means hold down the "Ctrl" key and press the "R" key simultaneously. **CTRL-A** means hold down the "Ctrl" key and press the "A" key simultaneously. Not all keyboards are the same; some keyboards may have a key labeled "Control" rather than "Ctrl."
- The keyboard sequence **ALT-TAB** in Microsoft Windows is used to toggle the active window. **ALT-TAB** means hold down the "Alt" key and press the "Tab" key simultaneously.

Nomenclature Used in This Manual

This manual uses many terms associated with Windows. Some of the terms are shown here:

Note: The restore button shown above changes depending on the size of the window. The graphic 🗗 is the restore button and means restore the window to its last dimensions, which were not full screen. The graphic 🗖 is the maximize button and means expand a window to occupy the entire screen.

vii

Contents

1. EDITING A BASIC SCHEMATIC 1
   1.A. Starting Multisim 1
   1.B. Placing Parts 4
   1.C. Correcting Mistakes 22
   1.D. Wiring Components 23
   1.E. Grounding Your Circuit 29
   1.F. Correcting Wiring Mistakes 33
   1.G. Labeling Nodes 33
   1.H. Displaying and Modifying the Title Block 35
   1.I. Problems 38

2. THE POSTPROCESSOR AND THE GRAPHER 43
   2.A. Creating a Plot with a Single Graph 46
   2.B. Creating a Plot with Two Graphs 53
   2.C. Creating a Plot with Three Graphs 59
   2.D. Saving and Loading Pages 66
   2.E. Deleting Items in the Postprocessor 70
      2.E.1. Deleting a Trace from a Graph 70
      2.E.2. Deleting a Graph from a Page 74
      2.E.3. Deleting a Page 77
   2.F. Modifying the Plot Properties in the Grapher 79
      2.F.1. Adding a Grid and Legend 80
      2.F.2. Trace Width, Color, and Label 82
      2.F.3. Changing Titles and Axis Labels 87
      2.F.4. Adding a Second Y-Axis 90
   2.G. Using the Cursors 95
   2.H. Zooming In and Out 101
   2.I. Saving and Opening Pages in the Grapher 110
   2.J. Problems 112

3. DC MEASUREMENTS 115
   3.A. Resistive Circuits 115
      3.A.1. Measurements with Indicators 115
      3.A.2. Measurements with the Multimeter 122
      3.A.3. Using the Power Meter 128
      3.A.4. DC Operating Point Analysis 132
   3.B. Nodal Analysis with Dependent Sources 136
   3.C. Diode DC Current and Voltage 140
      3.C.1. Changing the Temperature of the Simulation 144
   3.D. Finding the Thevenin and Norton Equivalents of a Circuit 146
   3.E. Transistor DC Operating Point 151
   3.F. Problems 162

4. DC SWEEP 167
   4.A. Basic DC Analysis 167
   4.B. Diode I-V Characteristic 180
   4.C. DC Transfer Curves 185
      4.C.1. Zener Clipping Circuit 185
      4.C.2. NMOS Inverter Transfer Curve 187

4.D. Nested DC Sweep — BJT Characteristic Curves	191
4.E. DC Current Gain of a BJT	194
4.E.1. H_{FE} Versus Emitter Current	194
4.E.2. H_{FE} Versus I_C for Different Values of V_{CE}	199
4.F. Problems	205

5. MAGNITUDE AND PHASE SIMULATIONS — 210

5.A. Magnitude and Phase Measurements at a Single Frequency	210
5.A.1. Magnitude Measurements with Instruments	210
5.A.2. Magnitude and Phase Measurements – AC Analysis	217
5.B. Bode Plots	230
5.B.1. Bode Plots Using the Bode Plotter Instrument	230
5.B.2. Bode Plots Using the AC Analysis	235
5.C. Amplifier Gain Analysis	243
5.D. Operational Amplifier Gain	247
5.E. Parameter Sweep — OPAMP Gain Bandwidth	259
5.F. AC Power and Power Factor Correction	265
5.F.1. Power Factor Correction	267
5.G. Measuring Impedance	269
5.G.1. Resistive Measurement with the Multimeter	269
5.G.2. Impedance Measurement of a Passive Circuit Using SPICE	271
5.G.3. Impedance Measurement of an Active Circuit Using SPICE	275
5.H. Problems	278

6. TIME DOMAIN ANALYSES — 292

6.A. Using the Oscilloscope Instrument	292
6.A.1. Timebase	295
6.A.2. Channel A and Channel B Volts per Division Settings	297
6.A.3. Trigger Settings	302
6.A.4. Using the Cursors	306
6.B. Phase Measurements of a Capacitive Circuit	309
6.C. Phase Measurements of an Inductive Circuit	313
6.D. Series LCR Resonant Circuit	318
6.E. Regulated DC Power Supply	327
6.E.1. Regulated DC Power Supply – Virtual Lab Simulation	331
6.E.2. Regulated DC Power Supply – SPICE Transient Simulation	338
6.F. Zener Clipping Circuit – SPICE Transient Analysis	354
6.G. Zener Clipping Circuit – Virtual Lab Simulation	361
6.G.1. Plotting Transfer Curves	365
6.H. Amplifier Voltage Swing	368
6.H.1. Fourier Analysis with Multisim	372
6.I. Ideal Operational Amplifier Integrator	379
6.J. Operational Amplifier Schmitt Trigger	385
6.K. Parameter Sweep — Inverter Switching Speed	390
6.L. Temperature Sweep — Linear Regulator	396
6.M. Problems	403

7. DIGITAL SIMULATIONS — 410

7.A. Digital Indicators, Signal Generators, and Instruments	410
7.A.1. Word Generator and the Logic Analyzer	410
7.A.2. Digital Probe and Undecoded Bar Graph Lights	423
7.A.3. Mixed Signal Indicators	427
7.B. Mixed Analog and Digital Simulations	432
7.C. Pure Digital Simulations	437

7.D. STARTUP CLEAR CIRCUIT	**440**
7.E. REAL AND IDEAL MODE DIGITAL SIMULATIONS AND GATE DELAYS	**443**
7.E.1. IDEAL MODE DIGITAL SIMULATIONS	443
7.E.2. REAL MODE DIGITAL SIMULATIONS	445
7.F. PROBLEMS	**448**
8. INDEX	**449**

PART 1
Editing a Basic Schematic

In this part we will cover the steps for drawing a circuit using Multisim. This includes locating and placing parts, wiring parts together, grounding your circuit, and labeling nodes. In Part 2, we will run a simple simulation and then discuss how to create plots using Multisim's graphic postprocessor. Parts 1 and 2 of this manual are introductory chapters that cover the skills necessary to create circuits and view results. The remaining chapters discuss how to run circuit simulations. You should cover Parts 1 and 2 in sequence. You can then cover examples in the remaining parts at random. Some of these examples will reference techniques covered in Parts 1 and 2.

1.A. Starting Multisim

If Multisim was installed properly on your computer, it can be easily started from the Windows **Start** menu. However, depending on how the Windows desktop is configured, and your version of Windows, the **Start** menu may appear differently. The screen captures shown here were generated using Windows XP Professional using the classic version of the **Start** menu. Usually, the **Start** menu is displayed at the bottom of the desktop:

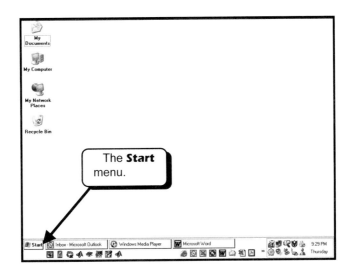

At this point, you could click on the **Start** button and continue:

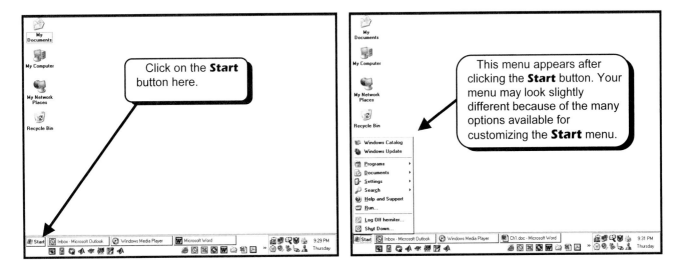

However, if the desktop appears as one of the screen captures shown below

1

the **Start** menu is hidden from view. There are two ways to make the **Start** menu appear. The first way is to bring the mouse pointer down to the bottom of the screen. After a moment, the **Start** menu should appear:

 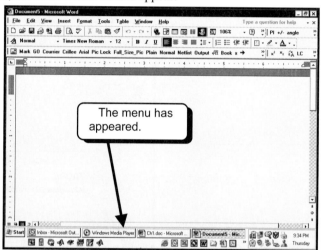

This method works if the **Start** menu is configured so that it is always on top and auto-hide mode is selected. At this point, you could click on the **Start** button and continue:

 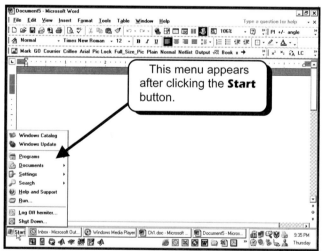

However, if you still cannot see the **Start** menu, you have three more options. (1) You can press the key on the keyboard or (2) type **CTRL-ESC**. These keys will bring the **Start** menu to the top and also select the **Start** button. (3) If you have an older keyboard and do not have the key, you can move the mouse to the bottom of the screen and drag the **Start** menu up. For this example, we will press the key:

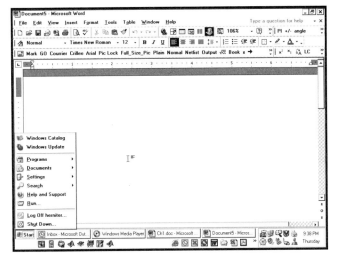

We now have the **Start** menu displayed. Select **Programs**:

After selecting **Programs**, the programs and program groups for your computer appear. The Multisim programs are contained in the group **Multisim 2001**. Click the *LEFT* mouse button on the text **Multisim 2001**. This will display the programs contained in this group:

Click the *LEFT* mouse button on the item **Multisim 2001** to run the Multisim circuit simulation package.

Multisim should run and display an empty schematic:

Before we start creating a schematic, we will enlarge the drawing paper. Click the **LEFT** mouse button on the Maximize button ▢ as shown below:

We now have a full window in which to create our schematic.

1.B. Placing Parts

Before we place any parts, we will check one of the Multisim options so that we are all using the same settings. From the Multisim menus, select **Options** and then **Preferences**:

Select the **Component Bin** tab:

Select the option **Place single component**:

With this option, we can only place a single component at a time and avoid the error of accidentally placing two identical components, one on top of the other. If you select the **Set as Default** button, the changes you have made to the **Preferences** will be used for all future schematics you create. If you do not select the **Set as Default** button, the changes will only apply to the current schematic. Any schematics you create in the future will use the previous option settings. Click on the **Set as Default** button and then click on the **OK** button to return to the schematic.

We will now create a schematic. The first part we will get is an independent voltage source. There are two methods for selecting a part: using the toolbars and component bins, and using the Multisim menus and a dialog box. We will first show how to use the toolbars and part bins. The parts toolbar is located on the left side of the Multisim window:

Editing a Basic Schematic

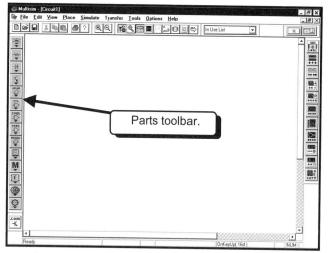

These buttons open up additional toolbars with parts in a specific category. An enlarged and rotated view of this toolbar is shown below:

If you hover the mouse pointer over a button, the name of the category will be displayed. For example, if we hover the mouse pointer over the [button], the text **Sources** will be displayed:

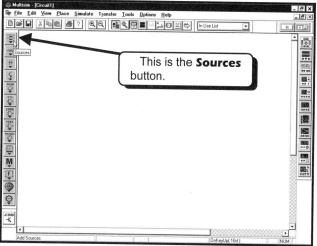

Looking at these buttons and using the hover feature, we see that there are categories of parts such as:

- Sources – DC and AC independent voltage and current sources, dependent sources, ground symbols, and non-linear sources.

- Basic – Resistors, capacitors, inductors, transformers, relays, switches, and more.

- Diodes – Diodes, Zeners, TRIACs, SCRs, LEDs, and diode bridges.

- Transistors – Bipolar junction transistors, MOSFETs, and JFETs.

- Analog – OPAMPs, comparators, and special function analog parts.

- TTL – TTL logic gates.

EWB Multisim   Editing a Basic Schematic   7

- CMOS – CMOS logic gates.

Not all parts can be obtained through these toolbars. For the parts that are available, the buttons are the fastest way to place them.

We will now place an AC voltage source in our circuit. Click on the **Sources** button to open the **Sources** parts bin:

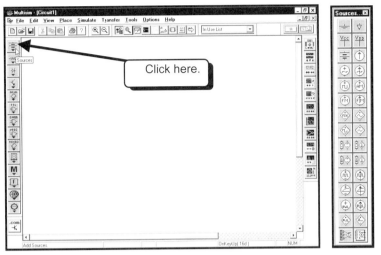

An enlarged and rotated view of the bin is shown below:

We see that this bin contains voltage and current sources, ground connections, dependent sources, and a few other sources that are not immediately obvious. To see the description of a button, hover the mouse pointer over a button. After a few moments, the description of the button will appear:

From the description and the graphic, it is usually easy to locate the part you are after. We want to place an AC voltage source, so click the *LEFT* mouse button on the **AC_VOLTAGE_SOURCE** button. The source will become attached to the mouse pointer and move with the mouse:

Move the source to the location where you want to place the part. Click the ***LEFT*** mouse button to place the part, and then move the mouse pointer.

We will now edit the properties of the source to make it a 12.6 V AC source with a frequency of 60 Hz. Before we do this, we will zoom in on the source. To zoom in or out, select **View** from the Multisim menus:

We will not use the menus to zoom in or out. However, the menus tell us which keys to press. We see that we can zoom in by pressing the **F8** key, and we can zoom out by pressing the **F9** key. The **F8** and **F9** keys have an advantage over using the menus or the toolbar buttons in that we can place the mouse pointer over the location in which we want to zoom and then press the **F8** or **F9** key. This allows us to zoom at a specific location. If you use the toolbars or the menus to zoom, it will zoom around some location, but not necessarily the area you want.

Place the mouse pointer over the AC voltage source and then press the **F8** key:

We cannot zoom in any further. If you want to zoom out and restore the previous magnification, press the **F9** key.

We will now edit the properties of this source. Double-click the ***LEFT*** mouse button on the graphic for the voltage source. The dialog box below will appear:

This portion of the dialog box allows us to change all of the physical parameters of the source such as the amplitude, frequency, and phase of the source. Note that we can specify the magnitude of the source as an amplitude or an RMS value. Both numbers control the size of the source; they are just a different way of specifying the same thing. Remember that the amplitude is the peak value of the sine wave. The relationship between the amplitude of a sine wave and the RMS value of a sine wave is:

$$RMS\ Value = \frac{Amplitude}{\sqrt{2}}$$

If you are working with RMS values in your circuit, you can specify the RMS value. If you are working with straight amplitude values, you can specify the amplitude. When you change one value, Multisim automatically changes the other.

When I said that I wanted to make the source a 12.6 V source at 60 Hz, I did not specify whether the number 12.6 was an RMS value or the peak value. We will specify 12.6 as the RMS value. Click the ***LEFT*** mouse button as shown to place the cursor in the **Voltage RMS** text field:

With the cursor in the text field, we can now modify its contents. Use the **BACKSPACE** key to erase the number that is already there and enter the number **12.6**. When you enter the number, the value in the **Voltage Amplitude** field automatically changes:

Note that $12.6 \times \sqrt{2}$ equals **17.82**. Thus, when we enter an RMS value, Multisim automatically calculates the amplitude for us. Had we entered an amplitude, Multisim would have automatically calculated the RMS value.

Also note that we can specify the units of the amplitude. By default, the units are in volts. However, we can change the units if we wish. Click the **LEFT** mouse button as shown to place the cursor in the units field for the amplitude:

When the cursor is in the units field, pressing the up or down arrow keys [↑][↓] on the keyboard will change the units. Several units are available including μV, mV, kV, and MV:

We want to create a 12.6 VRMS source, so leave the units as volts.

You can change the frequency to 60 Hz using the same technique:

To change the name of the voltage source, we need to select the **Label** tab:

The name of the source is already selected, so all we need to do is type the name we would like to use. We will change the name to Vx, so type the text **Vx**:

Click on the **OK** button to accept all of our changes. The changes will appear on the schematic:

Note that the amplitude of the source is displayed on the schematic, not the RMS value. You can leave the **Sources** toolbar open if you wish. I will close the toolbar to keep my screen from being cluttered. To close the toolbar, click the *LEFT* mouse button on the ☒ in the upper right corner of the toolbar:

We now need to place some resistors in the circuit. To place a resistor, click on the **Basic** button:

EWB Multisim　　　　　　　　　　Editing a Basic Schematic　　　　　　　　　　13

This is the toolbar that you will use to place most of the passive components such as resistors, capacitors, and inductors. An enlarged and rotated view is:

Note that there is a set of buttons with a green background, and the same buttons with a gray background. For example, the two resistor buttons are � and ▭. Either of these buttons will place a resistor. The green buttons are called virtual components. With a virtual component, you can specify any value for the component. For example, with a virtual resistor we could specify a resistance of 14758 Ω or with a virtual capacitor we could specify a capacitance of 1.37 µF. The virtual components do not specify a footprint for the component. A footprint is used to manufacture a PC board for your circuit.

The gray buttons are for selecting standard components. For example, if you select the gray resistor button ▭ you will be presented with a list of standard resistors that you could actually purchase. With standard resistors, you can only choose manufactured resistor values such as 1k, 1.2k, 1.5k, 1.8k, 2.2k, and so on. You could not specify a resistance of 14758 Ω. These resistors also specify a footprint for the component, so you can easily use the schematic for creating a PC board.

We will demonstrate using both standard and virtual resistors. For simulations, the two methods are equivalent. You can use whichever method you find easiest. We will first place a virtual resistor. Click the ***LEFT*** mouse button on the **Resistor_Virtual** button ▭ (the green one). A resistor graphic will become attached to the mouse pointer and move with the mouse:

Place the resistor as shown below and click the **LEFT** mouse button to place the part in your circuit:

The part is now in your circuit. We would like to make this resistor a 12.5 kΩ resistor and change its name to Rx. To change the properties of the resistor, double-click the **LEFT** mouse button on the resistor graphic:

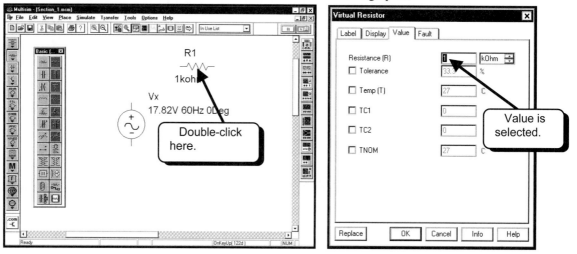

This screen lets us change the value of the resistor. We use the same procedure to change the resistance as we did to change the voltage of our source. Since the value of the resistor is highlighted, all we need to do is type in the new value. Type the text **12.5**. This will specify the value of the resistor as 12.5 kΩ because the units are selected as **kOhm**.

To change the name of the resistor, select the **Label** tab:

The present name of the resistor is **R1**. Since the name is highlighted, all we need to do is type the new name we want to use. We want to rename the resistor to Rx, so type the text **Rx**:

We are done changing the properties of this resistor, so click on the **OK** button. The changes will be displayed on the schematic:

If your values are not the same as shown, edit the properties of the device again.

Next, we need to place a 12 k resistor. We will demonstrate this using the standard resistor button. Click the **LEFT** mouse button on the gray **RESISTOR** button:

This dialog box lists most of the standard 5% resistor values you can purchase. You must scroll through this list and locate the resistor you need. Scroll down the list and find the one named **12kohm**. Click on the text to select it:

When you click on the **OK** button, the resistor will become attached to the mouse pointer and move with the mouse:

Place the resistor as shown and then click the *LEFT* mouse button to place the part in your circuit:

You can change the properties of this resistor using the same methods we showed for the virtual resistor. To edit this resistor's properties, double-click on the resistor graphic:

The one difference you will notice is that it is not obvious how to change the resistance value of the resistor. If you need to change the value, you should delete the resistor from your circuit and then place a new resistor with the correct value in your circuit. If you have a circuit where you know you will be changing the values of components, you should use the virtual components (the green buttons). The one advantage of using the standard components (the gray buttons) is that you guarantee that you are using a standard value that you can purchase.

With the above dialog box, you can still easily change the name of the resistor. Select the **Label** tab and change the name of the resistor to **Ry**:

Click on the **OK** button to accept the changes. The resistor will be renamed in the schematic:

Resistor **Ry** needs to be rotated. To rotate a component, click the *RIGHT* mouse button on the component. A menu will appear:

This menu shows several ways that we can rotate or flip the component. To accomplish our task, we can rotate the component either clockwise or counterclockwise. Also note that you can type **CTRL-R** or **SHIFT-CTRL-R** to rotate the part rather than use this menu. Select **90 Clockwise** to rotate the part:

The part is now oriented the way we would like.

One problem with our schematic is that we have zoomed in too far and we do not have enough room to add more components. You can zoom out by pressing the **F9** key, by selecting **View** and then **Zoom Out** from the menus, or by using the toolbars. We will show how to zoom out with the toolbars. Click the *LEFT* mouse button on the **Decrease zoom** button as shown below:

EWB Multisim   Editing a Basic Schematic   19

For the next resistor, we will show how to use the menus to place a part. This method is the most general method and can be used to place any part, including parts not available through the toolbar buttons. For example, this method is the only way to place a linear voltage regulator in your circuit. If you cannot find a part using the toolbars, you can use this method and search for your part. We will show how to place a resistor, but the method shown can be used to find any component. Select **Place** and then **Place Component** from the menus (or type **CTRL-W**):

The presently selected database is the **Multisim Master** database. You can select the User database, but this database contains no parts. The presently displayed **Component Family** is 2-volt CMOS components:

We want to place another resistor, so we need to select a different **Component Family**. Click the *LEFT* mouse button on the down triangle ▼ as shown to see the list of available component families:

Scroll down the list until you see a group that might contain the component in which you are interested:

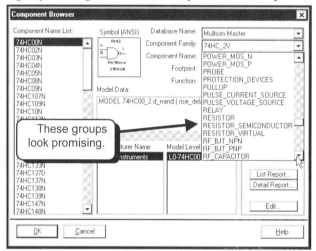

We see three possibilities, **RESISTOR**, **RESISTOR_SEMICONDUCTOR**, and **RESISTOR_VIRTUAL**. We will first try the **RESISTOR** group. Select **RESISTOR**:

Once we select a **Component Family**, all the parts in that family are displayed in the box. We can now select a part and place it in our circuit. Using this method, you can find any part in the database. The only confusion might come in finding the right **Component Family**.

The dialog box above lists all of the standard valued resistors. I want to place a resistor where I can easily change its resistance value. Thus, we need to select the **RESISTOR_VIRTUAL Component Family**:

We see that there is only one part in this family. This is because we can change the value of this resistor to any value, so only one part is needed. Select the part you want to use (it is already selected for us since there is only one part) and click on the **OK** button. The part will become attached to the mouse pointer and move with the mouse.

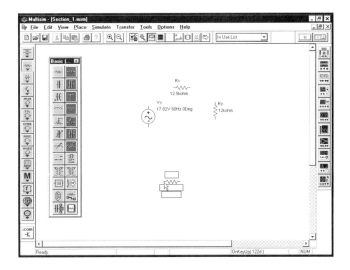

Use the same techniques to place the resistor as shown below and change its resistance value to 12 kΩ:

Use one of the methods shown previously to place a fourth resistor in your circuit.

Close the **Basic** toolbar by clicking on the ☒ in the toolbar's upper right corner.

Before I continue with the next section, I will move the components closer together and then press the **F8** key to zoom in. I am doing this to make the screen captures more readable. You do not need to do this if you are happy with the size of the components on your screen:

1.C. Correcting Mistakes

At this point you may have some mistakes in your schematic. To move parts follow the procedure below. For the moment we will assume that you wish to move a resistor.

1. Click the ***LEFT*** mouse button on the resistor graphic you wish to move. When the resistor graphic is selected, it will be surrounded by handles:

It may take several tries to highlight the resistor. Make sure that you select the resistor as shown above, and not its value or name as shown below:

2. When the resistor is highlighted, drag the resistor graphic to the desired spot in the schematic.

If you need to delete a part, follow Step 1 above. When the appropriate part is selected, press the **DELETE** key.

1.D. Wiring Components

We now must wire the components together. Before we do this we will change some of the wiring options in Multisim. Select **Options** and then **Preferences** from the Multisim menus:

Select the **Wiring** tab:

By default, **Autowire** is on. This causes Multisim to make some strange wire paths occasionally, so we will disable both **Autowire** options:

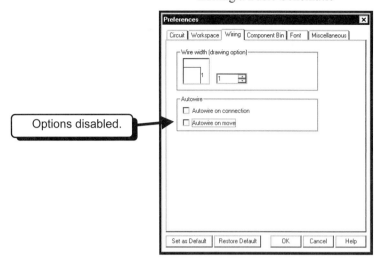

If you select the **Set as Default** button, the changes you have made to the **Preferences** will be used for all future schematics you create. If you do not select the **Set as Default** button, the changes will only apply to the current schematic. Any schematics you create in the future will use the previous option settings. Click on the **Set as Default** button and then click on the **OK** button to return to the schematic.

We are ready to wire the circuit. Place the mouse pointer over the top terminal of the voltage source. When you are near the pin the mouse pointer will be replaced by crosshairs:

When the crosshairs are displayed, you are ready to start drawing a wire. Click the *LEFT* mouse button to start a wire. As you move the mouse away you will see a dashed line connected to the crosshairs. As you move the mouse, the line grows and shrinks with the changing position:

If you made a mistake and do not really want to draw a wire, press the **ESC** key, and the dashed line will disappear. (Don't press the **ESC** key.) Move the crosshairs to point at the left terminal of **Rx**:

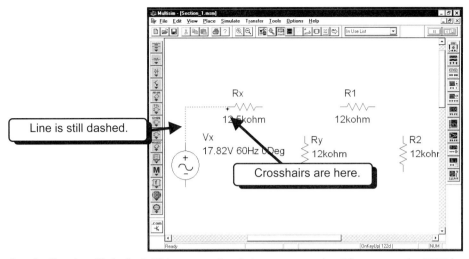

Note that the line is still dashed. This means that it is not yet a wire. If we press the ESC key, the dashed line will disappear and a wire will not be created. With the crosshairs placed as shown above, click the *LEFT* mouse button to create the wire. Note that the dashed line changes to a solid line, indicating that the line is now a wire connecting the two components. When you move the mouse pointer away, you will notice that the mouse pointer reappears and we are no longer drawing wires:

Use the same technique to connect **Rx** to **R1**:

Next, we need to connect **R1** to **R2**. Place the mouse pointer over the right terminal of R1 and click the *LEFT* mouse button to start drawing a wire. Move the mouse pointer directly to the right to draw a dashed line as shown:

With the crosshairs located as shown above, click the **LEFT** mouse button to create a corner. When you move the mouse down, a corner will be drawn:

Move the crosshairs to the top terminal of **R2** and click the **LEFT** mouse button to complete the wire:

Next, we need to connect **Ry** to the node between **Rx** and **R1**. Place the mouse pointer over the top terminal of **Ry** and click the **LEFT** mouse button to start drawing a dashed line:

Move the crosshairs so that they touch the wire between **Rx** and **R1**:

With the crosshairs as shown above, click the ***LEFT*** mouse button. The dashed line will become solid, indicating that it is a wire. When you move the mouse away, you will notice that the mouse pointer has returned and that there is a dot ┬ at the connection:

The dot indicates that there is a connection where two wires meet. If you do not see a dot, then the wires are not connected.

You should always use the method shown above for connecting one wire to another. **Do not use the procedure shown below**:

We first drew a corner and then connected a wire from **R1** to the corner. Occasionally this causes a wiring problem, so you should avoid using this method. Always use the method shown previously. When circuits do something strange during a simulation, suspect bad wiring. Delete all wires and then rewire your circuit.

We will continue by connecting the lower terminal of **Vx** to the lower terminal of **R2**. Use the techniques covered earlier:

We can now use the preferred method to connect **Ry** to the bottom wire:

Do not use the following method to wire components together:

Always try to make connections as a T. Avoid connecting to a corner or to a node.

1.E. Grounding Your Circuit

To simulate a circuit on Multisim, you must have at least one ground connection in your circuit. There are two ground symbols available with Multisim, the general ground symbol ($\perp$) and the digital ground symbol ($\triangledown$). The general ground symbol is used for all simulations except the real mode digital simulations. The general ground can be used for ideal mode digital simulations, which is the default mode for digital simulations. To see the difference between ideal and real mode digital simulations, see section 7.E. The general ground simulation works for all of the simulations covered in this text except the real mode digital simulations. When in doubt, use the general ground symbol ($\perp$). Both ground symbols are located in the **Sources** toolbar. To obtain the **Sources** toolbar, click the *LEFT* mouse button on the **Sources** button:

To place the ground symbol, click the **LEFT** mouse button on the **GROUND** button . The ground symbol will become attached to the mouse pointer and move with the mouse:

Locate the part as shown and click the **LEFT** mouse button to place the part:

I will close the **Sources** toolbar so that we can see the entire circuit:

We now need to wire the ground connection to our circuit. Place the mouse pointer over the pin for the ground symbol as shown. The mouse pointer will change to crosshairs:

Click the **LEFT** mouse button to start drawing a wire. As you move the mouse away, a dashed line will be drawn:

Move the mouse up to position the crosshairs on the wire as shown:

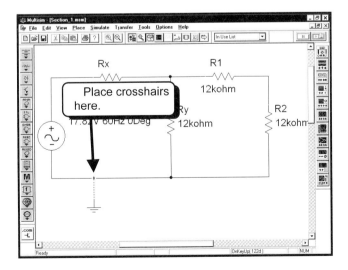

Click the **LEFT** mouse button to make the connection. The dashed line changes to a solid line, indicating that a wire has been created:

The wiring procedure above should be used rather than the two methods shown below:

Always try to make connections as a T. Avoid connecting to a corner or to a node.

Note that using the general ground symbol (⏚) specifies a node name of 0. If you double-click on a wire connected to the ground node, you will see that the name of the node is zero (**0**):

For SPICE simulations, the standard name for the ground node is zero, and all circuits must contain at least one node that is named zero (0). Note that the digital ground symbol (⏄) names the node VGND, not 0.

1.F. Correcting Wiring Mistakes

If you make a mistake when drawing a wire, you should use the following procedure to rework the wiring to the node to which the wire was connected:

1. Delete all wires connected to the node. To delete a wire:
   a. Move the mouse pointer to the segment of wire that you wish to remove.
   b. Click the **LEFT** mouse button on the wire you wish to remove. This will select the wire.
   c. Press the **DELETE** key to delete the selected wire.
2. Once all wires have been deleted, rewire all the components together using the preferred method shown in section 1.D.

1.G. Labeling Nodes

Multisim automatically numbers nodes for you. With SPICE simulations, you need to know the node names in order to display traces. Before we label any traces, we will select the option to display node names on the schematic. Select **Options** and then **Preferences** from the Multisim menus:

By default, node names are not displayed on the schematic. For this section, we would like to display node names, so enable the option and then click on the **OK** button:

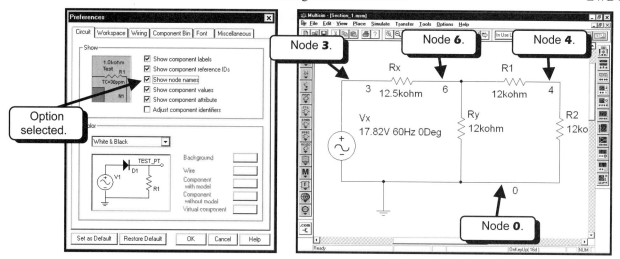

Node names are now displayed on the schematic. Note that your node names will be different than mine. Also note that if you rewire a section of your circuit, the node name of the node will change. The numbers are assigned sequentially, but you usually do not know a node name unless you give it your own name.

To rename a node, double-click the ***LEFT*** mouse button on a wire connected to the node. For example, we will name the left node Vin. Double-click the ***LEFT*** mouse button as shown below:

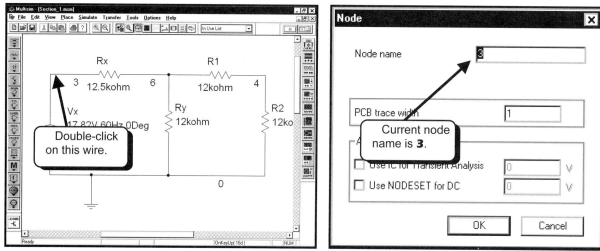

We see that the current name of the node is **3**. Since the name is highlighted, all we do to rename the node is type the new name. Type the text **Vin** and then click on the **OK** button. The new name will be displayed on the schematic:

The location of the node name is not so good. You can drag any of the text on the screen by clicking on the text once to select it. Once the text is selected, you can drag the text to a new location:

Using the same procedure, rename the center node **V_mid** and the right node **Vout**:

Naming nodes does not change any simulation properties. It just makes it easier for us to plot traces using the Postprocessor that is described in Part 2. **Node names should not contain any spaces.**

Now that we have finished the schematic, we need to save it. Select **File** and then **Save** from the menus, type **CTRL-S**, or click on the save button in the toolbars.

1.H. Displaying and Modifying the Title Block

The title block contains information about a project such as the name, who created it, who checked it, when it was last modified, the number of pages, and so on. For our example, we will just use it to identify the project, our name, and our class. To display the title block, select **Options** and then **Preferences** from the Multisim menus:

Select the **Workspace** tab:

By default, the title block is not displayed. Select the option and click on the **OK** button. The title block will be displayed, but we cannot see it with the current view:

The title block is located in the lower right corner of the schematic. If you zoom out (**F9**) and scroll the window to the lower right corner, you will see the title block:

To modify the contents of the title block, select **Options** and then **Modify Title Block** from the menus:

Fill in the title block fields with the information you want and click on the **OK** button:

The contents of the title block are:

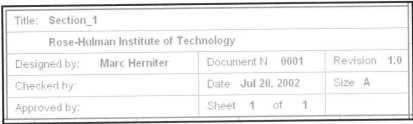

If you press the **F9** key to zoom out further, you will see how the title block is displayed on the schematic page:

We are now done with this drawing. We will be using this circuit in Part 2, so save your changes before continuing to Part 2 or exiting the program.

1.I. Problems

For the problems below, draw the specified circuit with the same parts. Use the same part numbering and give the nodes the same names if specified. Turn in a schematic that includes a title block with your name, class, and problems specified in the title block. Most of the circuits contain components that we have not yet used. You should be able to find these parts using the parts bins.

Problem 1-1:

Problem 1-2:

Problem 1-3:

Problem 1-4:

Problem 1-5:

Problem 1-6:

EWB Multisim — Editing a Basic Schematic

Problem 1-7:

Problem 1-10:

Problem 1-8:

Problem 1-11:

Problem 1-9:

Problem 1-12:

Problem 1-13:

Problem 1-14:

Problem 1-15:

Problem 1-16:

Problem 1-17:

Problem 1-18:

PART 2
The Postprocessor and the Grapher

The Postprocessor and the Grapher are two Multisim programs that allow you to display the results obtained from a SPICE simulation graphically. We will be using these programs extensively throughout this manual to display the results of simulations. Various aspects of the Postprocessor and Grapher are discussed in sections throughout this manual. However, if you pick specific examples you may miss those showing how to use some of the tools provided by the Postprocessor and Grapher. The result may be that a section in the manual you are currently using refers to a tool that was discussed in a section that you did not cover. To avoid this problem, we will review the most frequently used tools in the Postprocessor and the Grapher.

This entire chapter is based on the graphs we create in sections 2.A through 2.C. It is best to do this entire chapter in one sitting because the sections after section 2.C use the graphs created in the first three sections. However, if you are not able to do this, you should cover sections 2.A through 2.C, and then finish by doing section 2.D. This section shows you how to save and restore the graphs created in the first three sections. Using the methods in section 2.D, you can easily re-create your work in a few steps and then continue with the later sections.

To demonstrate the Postprocessor and Grapher, we will simulate the circuit drawn in Part 1 with a Transient Analysis:

One important note about the schematic above is that we have renamed the nodes **Vin**, **V_mid**, and **Vout**. We have also selected to display the node names on the schematic. We need to know the node names in order to use the Postprocessor. See section 1.G for instructions on naming nodes and displaying node names on the schematic. Assuming that you have followed and completed Part 1, we will start by setting up the Transient Analysis. Select **Simulate**, **Analyses**, and then **Transient Analysis** from the Multisim menus:

Change the parameters indicated below:

43

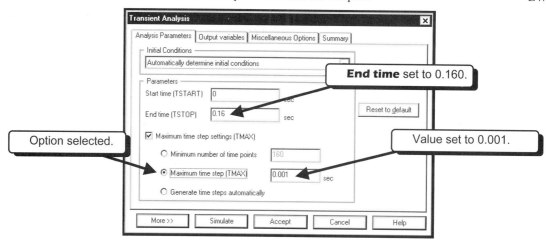

These options are discussed in detail in Part 6 and will not be discussed here.

Whenever you run a SPICE simulation, whether it is a Transient Analysis, AC Sweep, DC Sweep, or one of the others, you must specify the **Output variables** for that simulation. The **Output variables** you select are the only ones that can be plotted by the Postprocessor and the Grapher. If you have a circuit with 100 nodes, only the data from the nodes you specify as **Output variables** will be accessible to the Postprocessor and Grapher. Select the **Output variables** tab:

We see that the three node voltages (**v_mid**, **vin**, and **vout**) are available as selections. The item vvx#branch is the current through voltage source Vx. In a typical simulation, you would only select the items that you are interested in. For this example, we will select all of the variables listed. You can add one variable at a time, or add several at one time. To add a single item, click the *LEFT* mouse button on the item to select it, For example, click the *LEFT* mouse button on the text **v_mid** to select the item:

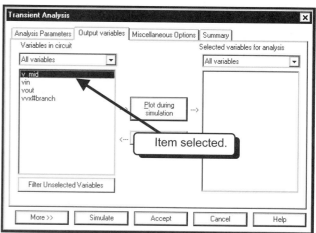

Click on the **Plot during simulation** button to make the voltage data available for that node to the Postprocessor:

Items listed in the right pane are available to the Postprocessor and may be plotted when the simulation is complete.

We would like data for all voltages and currents to be available for our example, so we need to select more of the items in the left pane. To select multiple items, hold down the **CTRL** key and then click the *LEFT* mouse button on the items you want to select:

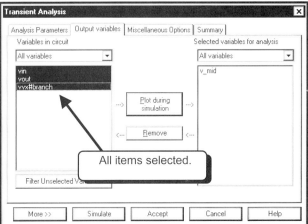

Click on the **Plot during simulation** button to make the data available for the selected items:

We are now ready to run the simulation, so click on the **Simulate** button. While the simulation is running, the Grapher will open and display a plot with all of the **Output variables** on the same plot:

Every time you run a simulation, a plot of all the output variables is automatically generated. This plot may or may not be useful depending on how you selected the output variables. Each time you run a simulation, a new Grapher page will be generated and displayed in the Grapher, whether you want that plot or not. In a small circuit, you may only need to plot one or two items. This feature then becomes a quick way of generating a plot of the variables in which you are interested. If you are simulating a large circuit and are interested in many voltages and currents, the automatic plot will not be too useful because it plots everything on a single graph. In our graph, the current trace is useless because it is so small compared to the magnitudes of the voltages that the current looks like a straight line, which it is not. We will now use the Postprocessor to create a new plot.

2.A. Creating a Plot with a Single Graph

We will now create a plot of the input voltage. Select **Simulate** and then **Postprocess** to open the Postprocessor:

We need to create a new page within the Grapher window. Click on the **New Page** button:

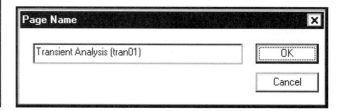

Since the data we are using came from a Transient Analysis, and this was the first Transient Analysis of this Multisim session, the default name of the page is **Transient Analysis (tran01)**. We would like to change the name to something more useful. Change the text to `Section 2 - Single Graph` and then click on the **OK** button:

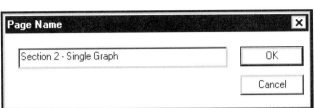

Notice that a tab appears and displays the name of the page we created. Each page we created will have a separate tab, and we can use the tabs to select which page we want to work with. For now, we only have a single page.

Next, we need to create a graph on this page. Click on the **New Graph** button:

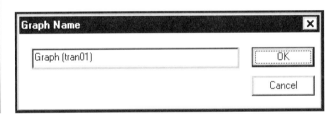

Note that the **New Chart** button creates text output instead of a graph. In a few examples, we will use this button to place text on one of the pages. Most of the time we will create graphs, as in this example. Multisim suggests a name for the graph. This will be the title of the plot. We are going to plot a few of the circuit voltages, so name the graph `System Voltages` and then click on the **OK** button

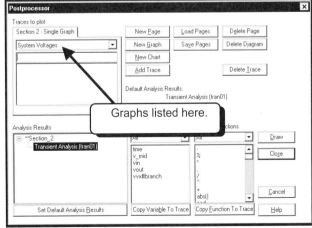

The newly created graph is listed in the window. Note that a page can have several graphs. For this example, we will only have a single graph on the page.

We are now ready to add a trace to the graph. We will first plot the input voltage. Click the *LEFT* mouse button on the text **vin** to select the text:

 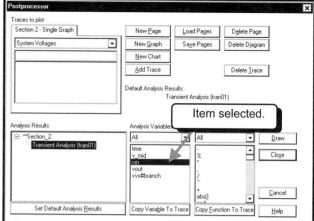

Next, click on the **Copy Variable To Trace** button:

 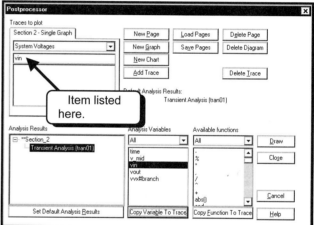

Next, click on the **Add Trace** button to create the trace:

 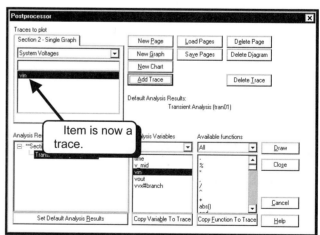

The item is now a trace that will be plotted on the graph named **System Voltages** on the page named **Section 2 – Single Graph**. At this point we could add more traces if we wanted to. However, we will now generate the plot so that we can see what we have created thus far. To create the plot, click on the **Draw** button:

After clicking on the **Draw** button, you may or may not have noticed anything happen. The new page is displayed in the Grapher, but the Grapher is not the top window. To display the Grapher we must close the Postprocessor window. Click on the **Close** button:

If you do not see the Grapher, type **CTRL-G** or select **View** and then **Show Grapher** from the Multisim menus. We would like to enlarge the Grapher to occupy the full window. Click the **LEFT** mouse button on the maximize icon □ as shown:

Note that the Grapher has two tabs, one for each plot (or page created by the Postprocessor). Right now we have two pages: the first was created automatically when we ran the simulation, and we generated the second using the Postprocessor. You can display either page by selecting the tab for the page:

We will show how to customize graphs in later sections. For now we are just adding traces and creating pages. We will now add a second trace to this graph. We need to switch back to the Postprocessor, so select **Simulate** and then **Postprocess** from the Multisim menus:

 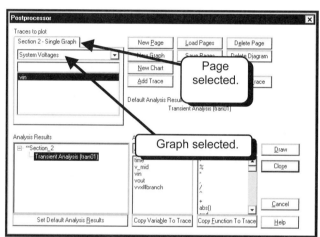

Since we only created a single page, and there is only one graph on that page, the one we are interested in is already selected. Later on, after we have created several pages and several graphs on a single page, we will need to select the page and graph before we proceed. Right now, any traces we add will be added to the graph we just created.

We would like to add the **vout** trace to out plot. We will use a slightly different method this time. Double-click the *LEFT* mouse button on the text **vout**. This will automatically copy the item to the trace:

 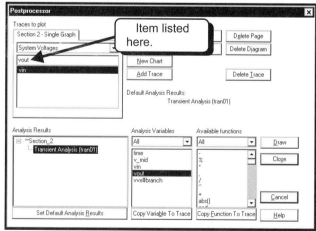

Click on the **Add Trace** button to create the trace:

The item is now a trace. We note that this graph will now have two traces, **vin** and **vout.** To create the plot, click on the **Draw** button:

A new page is displayed in the Grapher, but the Grapher is not the top window. To display the Grapher we must close the Postprocessor window. Click on the **Close** button.

If you do not see the Grapher, type **CTRL-G** or select **View** and then **Show Grapher** from the Multisim menus.

There is an error in one of the traces:

We see that when the red trace reaches the end of the time axis, a straight line is drawn back across the graph. This is an error in the Grapher that will be corrected in a future release. The trace is correct except for the straight line. You can avoid this error by creating multiple graphs on a single page, and then adding only one trace per graph.

Another thing we notice about the Grapher is that there are now three tabs. When we clicked on the **Draw** button, the first page we created was not modified. Instead, a new page was created which contained all of the changes. If you select the middle tab, you will see the first page we created:

If you want to keep this page, then you don't need to do anything. If you want to delete the page, click the ***LEFT*** mouse button on the cut button as shown. The page will be deleted:

The page is deleted and only two tabs are left. The method shown can be used to delete unwanted pages in the Grapher. When you run several simulations, the Grapher can become filled with old plots that are no longer needed. Use this method to remove the unwanted plots.

You can add more traces to the graph as you want. Below, I have added the v_mid trace to the graph:

We see that a new tab was added, so we can delete the older plot if we want. Also, at the end of the blue trace is a straight line. This is the same error that we saw earlier. The sinusoidal part of the trace is correct. The only part that is incorrect is the straight line at the end of the trace. Delete the tab for the unwanted page:

2.B. Creating a Plot with Two Graphs

For our next example, we will create a new Grapher page that has two graphs on it. Open the Postprocessor by selecting **Simulate** and then **Postprocess** from the menus:

Click on the **New Page** button to create a new page in the Grapher. Enter a name for the page, such as `Section 2 - Two Graphs`:

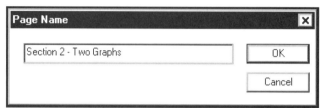

When you click on the **OK** button, a new tab will be added to the dialog box:

The Postprocessor dialog box now has two tabs. The top tab is the page we just created, and it is also the page we are working with in the Postprocessor. If you want to modify other pages, you can always select a different tab. We want to work with this newly created page, so we will leave the tabs as they are. We now want to create a new graph on this page, so click on the **New Graph** button, and enter a name for the graph. On this page I am going to plot the input voltage and the current through the source Vx. This first graph will be the current, so name the graph `Source Current`:

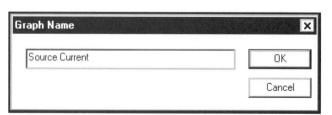

When you click on the **OK** button, the graph **Source Current** will be added to the list of graphs for this page:

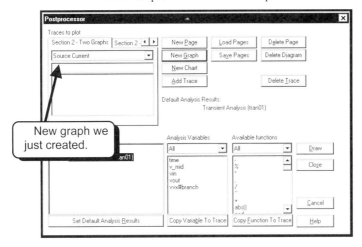

Any traces we add will be placed on the graph named **Source Current**, which is on the page named **Section 2 – Two Graphs**. We will now add a trace to this graph. Double-click on the text **vvx#branch** to select the variable:

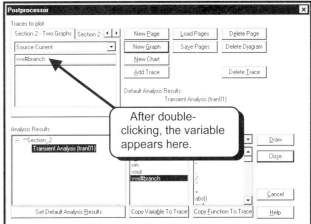

Click on the **Add Trace** button to plot this trace on the graph:

The branch current is now a trace on the graph.

We would now like to create a second graph on this page. Click on the **New Graph** button to create a second graph on the selected page and enter a name for the graph, such as `Input Voltage`:

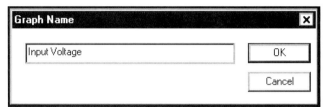

When you click on the **OK** button, the new graph will be created:

It may not be obvious that there are two graphs on this page. To see the plots on the page, click the **LEFT** mouse button on the down triangle ▼ as shown:

We see that there are two graphs listed here named **Source Current** and **Input Voltage**. This list allows us to select which graph we want to work with. Select the **Input Voltage** graph before continuing. It was selected before we clicked on the down triangle ▼, but make sure it is still selected after looking at the available graphs:

Next, we would like to plot the input voltage (**vin**) on this graph. Double-click the *LEFT* mouse button on the text **vin**:

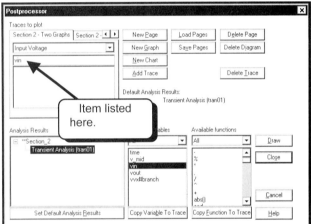

Click on the **Add Trace** button to make **vin** a trace on the graph named **Input Voltage** on the page named **Section 2 – Two Graphs**:

We are now done with these graphs and need to draw the page. Click on the **Draw** button to create the page in the Grapher:

The new page was created, but we cannot see it with the Postprocessor window open. Click on the **Close** button to close the Postprocessor:

The page looks as expected. One thing we note about a branch current is that it is positive when it flows into the positive voltage terminal of the source. Since our voltage source supplies current to our circuit, the current is actually leaving the positive terminal when the voltage is positive. This is why the current is shown as negative when the voltage is positive.

Another thing we notice is that every time we click on the Draw button, it creates a new page for every page listed in the Postprocessor. Our Postprocessor had two pages defined. When we clicked on the Draw button, both pages were added to the Grapher window, not just the page we were working on. In our example, this had the effect of creating a second copy of the page titled **Section 2 – Single Graph**, because this page was already displayed in the Grapher from the previous example. Thus, if you define multiple pages in the Postprocessor, all of the pages will be created as new pages in the Grapher each time you click on the Draw button. In our example, this has the effect of creating several unwanted pages in the Grapher. To fix this problem in our example, we will delete the unwanted pages. Click the *LEFT* mouse button on one of the tabs labeled as **Section 2 - Single Graph**:

Click on the **Cut** button to delete the selected page:

The page was deleted. You will need to use this technique to clear out many of the plots that appear in the Grapher.

2.C. Creating a Plot with Three Graphs

As a last example, we will create a new page with three graphs on it. On this page we will show some of the math functions available in the Postprocessor. Select **Simulate** and then **Postprocess** from the menus to open the Postprocessor, then click on the **New Page** button and give the page a name:

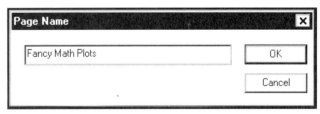

Click on the **OK** button to create the new page:

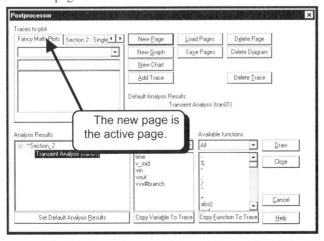

We see that the new page we just created is also the active page. Everything we do will be done to the active page.

We now want to create three graphs on this page. Click on the **New Graph** button, name the graph **Resistor Power**, and click on the **OK** button:

Click on the **New Graph** button, name the graph `Source Power,` and click on the **OK** button:

Click on the **New Graph** button, name the graph `Input Voltage and Current,` and click on the **OK** button:

The currently selected graph is the one we just created, named **Input Voltage and Current**. We will now add the input voltage and the source current trace to this plot. Double-click on the text **vin** to select the variable, and then click on the **Add Trace** button to create the trace:

We would like to plot the source current on the same plot. The only problem is that the current is about 10,000 times smaller than the source voltage in magnitude. In order to be able to see the current trace on the same set of axes as the voltage trace, we need to multiply the current trace by 10,000. (We can also solve this problem by using a different y-axis for the current trace. This method is covered in section 2.F.4.) We need to add the trace 10000*vvx#branch. We will manually enter the text 10000*. Click the *LEFT* mouse button as shown to place the cursor in the text field:

With the cursor located as shown above, type the text **10000***:

Next, double-click the *LEFT* mouse button on the text **vvx#branch** to add the text to the field:

Click on the **Add Trace** button to add the trace to the graph:

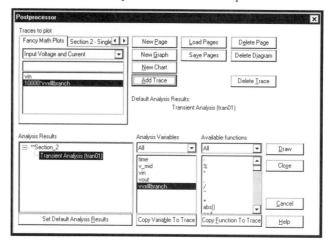

The graph named **Input Voltage and Current** now has two traces plotted on it.

Next, we would like to work with the graph named Source Power. Click the *LEFT* mouse button on the down triangle ▼ as shown to view the list of graphs on the page:

 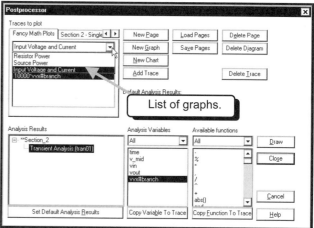

Select the graph named **Source Power**:

Create the text `vin*vvx#branch` and add it as a trace for the selected plot:

The last thing we want to do is add two traces to the Resistor Power graph. Select the graph named Resistor Power:

We would like to plot the average and instantaneous power absorbed by resistor Rx. The instantaneous power is the current through Rx times the voltage across Rx. This is equal to $-vvx\#branch*(Vin-V\_mid)$. Add this trace. You can type it out, or use a combination of double-clicking the mouse on the variable text and typing the symbols:

We will add a second trace to this plot, which calculates the average power absorbed by the resistor. Multisim does not have a time average function; however, it does have a mean function. This function calculates the mean of all of the elements of a trace. The result of the mean function is a single number that is the mean over all time. This is not the same thing as a running time average, but it can be used as an approximation for calculating the time average power. Since the function returns a single value, it will be plotted as a straight line.

To see the available functions, click the **LEFT** mouse button on the down triangle ▼ as shown:

Some of the functions are obvious, others are not.* Scroll down the list until you see the text **mean()**. Double-click on this text to add it to the text field:

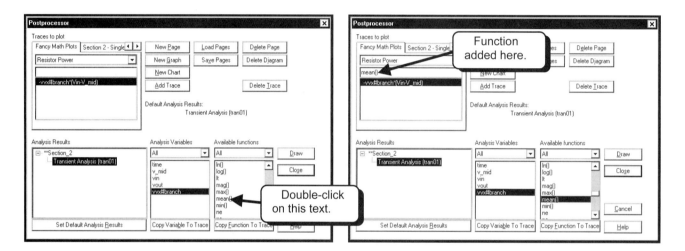

Use the mouse to place the cursor between the parentheses:

With the cursor located between the parentheses, you can create the remainder of the text string using the techniques covered earlier:

---

* To see a list of the functions and their properties, click on the **Help** button.

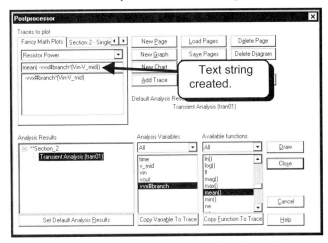

Click on the **Add Trace** button to make the text string a trace on the selected graph:

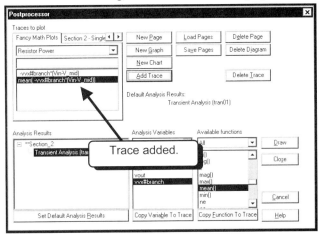

We are now done with all of the graphs, so click on the **Draw** button to create the pages. Then close the Postprocessor and view the plots with the Grapher.

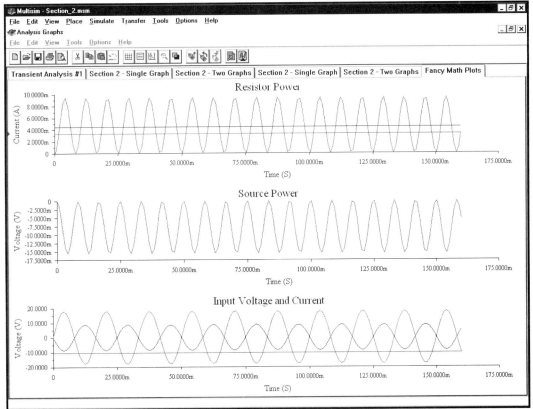

If you are using a Windows screen resolution of 800 by 600 or less, you will not see the titles **Resistor Power**, **Source Power**, and **Input Voltage and Current** on your page because there is not enough room to display the titles. Also note that when we clicked on the Draw button, three plots were added to the Grapher window. We will again need to delete some of the duplicate tabs.

Before stopping and exiting Multisim, you may want to complete the next section because it shows you how to save and recover all the work we just did. After completing the next section, you can stop and then complete the remaining sections at a later time.

2.D. Saving and Loading Pages

We have spent a great deal of time creating graphs that describe the function of our circuit. We would like to save these plots for future use or to send to someone so that they can view the results using the graphs we created. This can be easily done by saving and loading pages in the Postprocessor. Once you have created all of the pages and graphs you need, open the Postprocessor by selecting **Simulate** and then **Postprocess** from the Multisim menus:

Click on the **Save Pages** button:

Note that the initial directory is the same directory where the circuit files are saved. Specify a directory where you would like to save the file and then specify a name for the file. I will give the file the same name as the circuit file, `Section 2`:

Multisim automatically adds the extension .pgf, so the file will be named Section 2.pgf. The circuit file was named Section 2.msm, so you will not delete the circuit file. Click on the **Save** button to save the file. You will return to the Postprocessor. Close the Postprocessor, close the Grapher, and then save the circuit by selecting **File** and then **Save** from the Multisim menus:

We have now saved the Postprocessor pages and the circuit, so we can exit Multisim. At a later time, we can easily regenerate all of the simulation results with a few steps. Select **File** and then **Exit** to close down Multisim.

Now that we have saved the information, we will show how to restore the graphs from the previous sections. Open Multisim. You will start with a blank schematic:

We must first open the circuit file. Select **File**, **Recent Files**, and then the name of the circuit file we were just using. In my example, I called the file **Section 2.msm**:

You can enlarge the schematic window and center the circuit if you wish.

We must now rerun the simulation to generate the data for the circuit. We must repeat the simulation we ran earlier. Select **Simulate**, **Analyses**, and then **Transient Analysis** from the menus:

We must verify that the **Analysis Parameters** and the **Output variables** tabs of the dialog box are the same as before. They should be the same as before, because this information is saved with the circuit. However, it never hurts to check and make sure the information is correct:

Both windows look correct, so click on the **Simulate** button to rerun the simulation:

We must now open the Postprocessor and load the file that describes the pages we created. Select **Simulate** and then **Postprocess** from the menus:

Click on the **Load Pages** button:

Select the file we created earlier and click on the **Open** button:

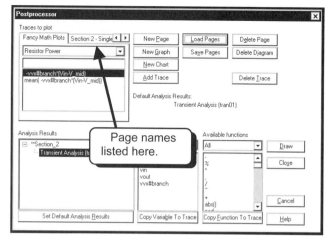

We note that the page names listed in the tabs shown above are the pages we created in sections 2.A to 2.C.

The last thing we must do is display the pages with the Grapher. Click on the **Draw** button to create the pages with the Grapher, and then click on the **Close** button to close the Postprocessor. When you look at the Grapher, you will see all the pages we created earlier:

2.E. Deleting Items in the Postprocessor

Now that we have spent a great deal of time creating Postprocessor pages and graphs and traces, we will show you how to delete them.

2.E.1. Deleting a Trace from a Graph

We will first show how to delete a trace from a graph on a specific page. We will show how to delete the current trace on the graph labeled Input Voltage and Current on the page named Fancy Math Plots. Before we show how to delete a trace, we will empty the Grapher window to eliminate any confusion when new pages are drawn. If the Postprocessor is open, close it. If the Grapher window is not open, type **CTRL-G** to open the Grapher window:

We want to clear all of these plots, so select **Edit** and then **Clear Pages** from the Grapher menus:

The default is to clear all of the pages, so click on the **Delete** button:

Now when we create new plots, it will be easy to identify them. Had we not cleared all the pages, we would have old plots and newly created plots, and there could be some confusion as to whether we are looking at old or new plots, and if the changes we made had any effect.

We must now open the Postprocessor and select the correct page, graph, and trace. Open the Postprocessor. In my window, the correct page is not displayed:

Use the mouse and the arrows to scroll the tabs until you see the tab named **Fancy Math Plots**:

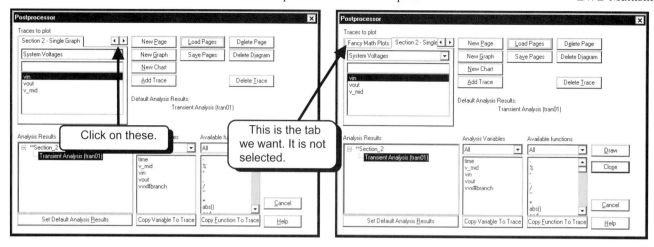

We see the tab we want, but it is not selected. Click on the tab to select it:

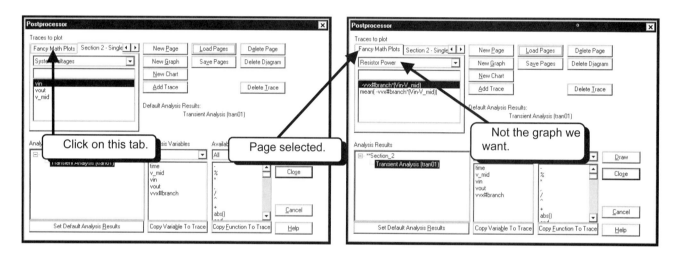

We have now selected the page named **Fancy Math Plots**. This page has three graphs on it, and we want to modify the graph named Input Voltage and Current. This graph is not yet selected. Click the *LEFT* mouse button on the down triangle ▼ as shown to view the list of graphs on this page:

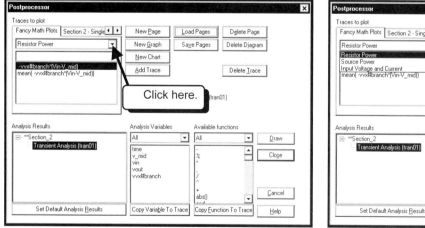

Select the graph named **Input Voltage and Current**:

EWB Multisim   The Postprocessor and the Grapher   73

This graph has two traces on it. We want to delete the current trace, so click the **LEFT** mouse button on the text **1000*vvx#branch** to select it:

 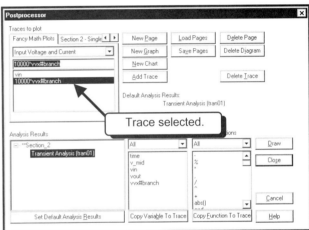

To delete the selected trace, click on the **Delete Trace** button:

 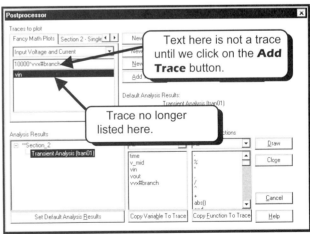

The trace has been deleted from the graph. To see the effects of our change, we must click on the **Draw** button to redraw all the Postprocessor pages. This will duplicate some earlier plots, but the newest version of the page named **Fancy Math Plots** will have the changes we just made:

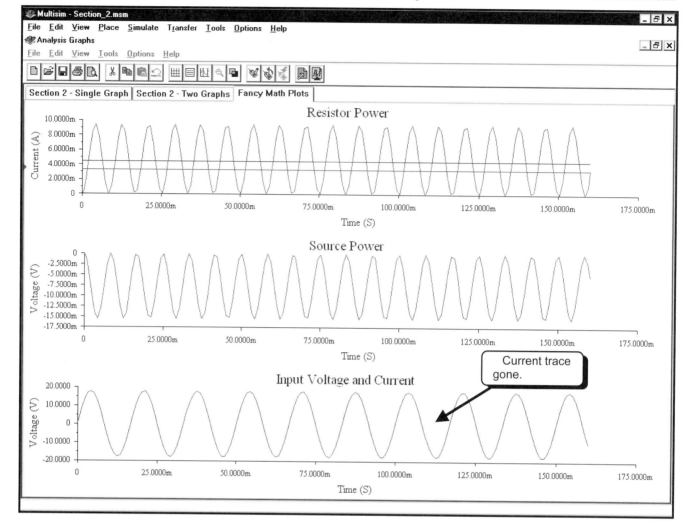

2.E.2. Deleting a Graph from a Page

We will now delete a graph from a page. Before we continue, delete all pages from the Grapher using the procedure shown at the beginning of section 2.E.1 on pages 70-71:

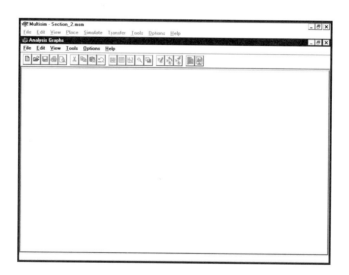

We would like to delete the graph named Input Voltage on the page named Section 2 – Two Graphs. Open the Postprocessor:

EWB Multisim · The Postprocessor and the Grapher · 75

On my Postprocessor window, the page **Fancy Math Plots** is still selected from the previous example. However, we can see the plot named **Section 2 – Two Graphs**, so all we need to do is select it. If you do not see the tab named **Section 2 – Two Graphs**, use the arrows to locate it. Click the *LEFT* mouse button on the tab labeled **Section 2 – Two Graphs** to select the page:

 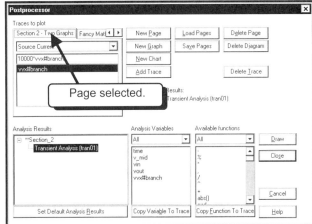

We must now select the graph named Input Voltage. To see the list of graphs for this page, click on the down triangle ▼ as shown:

Select the graph named **Input Voltage**:

We have now selected the graph we want to delete, so click on the **Delete Diagram** button:

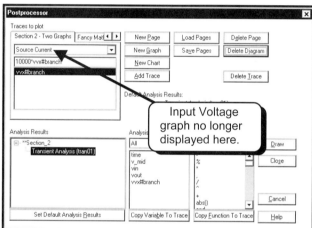

We see that the Input Voltage graph is no longer displayed in the text field. If we click on the down triangle to display the list of graphs, we will see that the Input Voltage graph is no longer listed:

We see that this Postprocessor page only has a single graph now.

To see the results of the changes we have made, click on the **Draw** button. Close the Postprocessor and open the Grapher. Select the page named Section 2 – Two Graphs:

EWB Multisim   The Postprocessor and the Grapher   77

We see that this page now only contains a single graph.

2.E.3. Deleting a Page

We will now delete an entire page. Before we continue, delete all pages from the Grapher using the procedure shown at the beginning of section 2.E.1 on pages 70-71:

We would like to delete the page named Section 2 – Single Graph. Open the Postprocessor. In my window, the correct page is not displayed:

Use the mouse and the arrows to scroll the tabs until you see the tab named **Section 2 – Single Graph**:

We see the tab we want, but it is not selected. Click on the tab to select it:

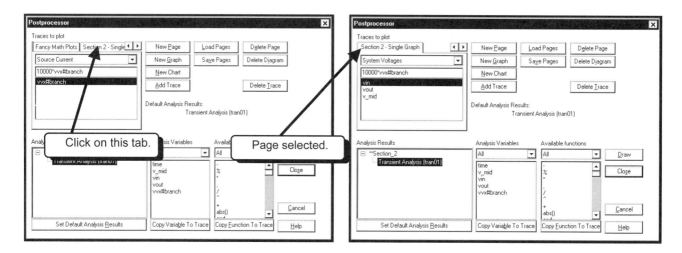

We have now selected the page we want to delete. Make sure that you always click on the page tab to select the page or you may delete the wrong page. To delete the page, click on the **Delete Page** button:

The page is deleted. If you use the arrows to scroll the tabs, you will notice that there are now only two pages (tabs):

EWB Multisim   The Postprocessor and the Grapher   79

To see the results of the changes we have made, click on the **Draw** button. Close the Postprocessor and open the Grapher. The Grapher will now only have two tabs, indicating only two pages. The two pages are shown below:

No graph titles are shown on the left plot because this is an 800 × 600 resolution screen capture. In earlier screen captures of the page with three plots, the screen captures were taken at 1024 × 768 resolution and these plots displayed the titles of each graph.

2.F. Modifying the Plot Properties in the Grapher

The Postprocessor allows us to create graphs and place traces on those graphs. The Grapher allows us to customize the graphs. This includes adding a grid and legend, modifying trace widths, changing labels, and changing the axes scales. We will continue with the pages generated in the earlier sections. If you do not have these pages displayed, follow section 2.D to load the plots. We will use the Grapher page with a single plot:

2.F.1. Adding a Grid and Legend

There are two ways to add a legend and a grid to a graph. We will show both techniques here. On the page shown above, click the **RIGHT** mouse button on an empty space in the graph. A menu will appear:

To turn on the grid, select **Toggle Grid** from the menus:

 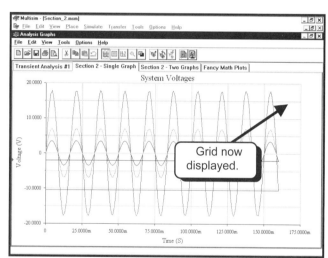

To display the legend, repeat the above procedure, but select **Toggle Legend**:

If you want to turn off the grid or the legend, just repeat the above steps.

There is a second way to turn on or off the display of the grid and the legend. For the second example, we will use the page that contains two graphs:

Each graph is modified separately. Thus, we will change the grid and the legend of one of the graphs. Click the **RIGHT** mouse button on an empty space in the bottom graph. A menu will appear:

Select **Properties**:

This dialog box allows us to modify all settings for a graph. For now, we will only change the display of the grid and the legend. On my dialog box the **General** tab is selected. If you do not have the **General** tab selected, then click on the tab to display it. Note that this portion of the dialog box allows you to display both the grid and the legend. By default, they are not displayed:

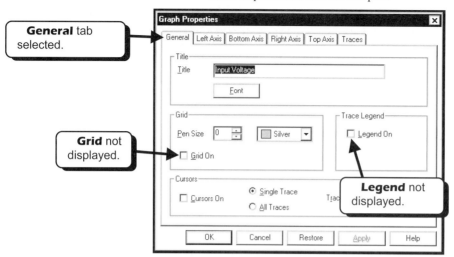

To display the **Grid** and the **Legend**, click on the square ☐ next to the options to place a checkmark in the box ☑:

When you click on the **OK** button, the grid and the legend will be displayed on the bottom graph only:

To display the grid and the legend on the top graph, you will need to repeat the procedure for the top graph. To obtain the properties dialog box of the top graph, you will need to click the ***RIGHT*** mouse button on an empty space in the top graph.

2.F.2. Trace Width, Color, and Label

For this example, we will work with the page that has a single plot. Click the ***RIGHT*** mouse button on an empty space in the graph. A menu will appear:

The Postprocessor and the Grapher

Select **Properties**:

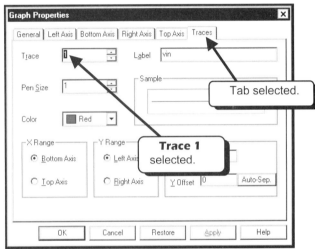

On my dialog box, the **Traces** tab is selected. This is the tab we need to use. Select the **Traces** tab if it is not selected.

By default, **Trace 1** is selected, so we will modify the properties of this trace first. We will first change the width of the trace. Click the *LEFT* mouse button as shown below to place the cursor in the **Pen Size** field:

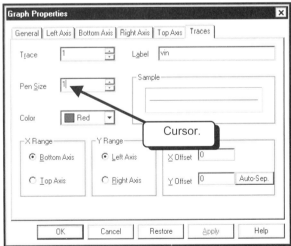

We can use the up and down arrows (↑↓) on the keyboard to change the value. We would like to change the thickness to 2, so press the up arrow key ↑ once:

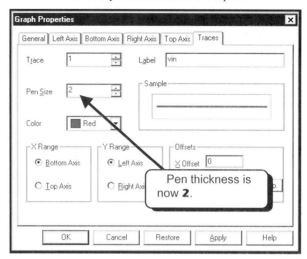

Next, we will change the color of the trace. Click the **LEFT** mouse button on the down triangle ▼ as shown:

Select the color you want to use. I will select **Purple**:

The last thing we will do is modify the label for this trace. Click the **LEFT** mouse button as shown to place the cursor in the **Label** text field:

We can now modify the text to suit our needs. Use the **BACKSPACE** key to erase the text and then enter the text `Input Voltage (Vin)`:

We are now done with this trace.

This graph has three traces, and we can modify the properties of each trace independently. Presently **Trace 1** is selected. To switch to another trace, click the *LEFT* mouse button as shown below to place the cursor in the **Trace** field:

We can now use the up and down arrows (⬆⬇) on the keyboard to change the value. We would like to change to trace 2, so press the up arrow key ⬆ once:

We are now working with **Trace** number **2**. We note that the properties of this trace are completely different than thoes for Trace 1. We can use the techniques shown earlier to modify the properties of this trace. Change the **Pen Size** to 3, the color to gray, and the **Label** to `Output Voltage (Vout)`:

Next, modify the properties of trace 3. Change the color to Red, the thickness to 4, and the label to **Middle**:

To see the changes and return to the Grapher, click on the **OK** button:

You can decide on the best trace color and thickness for your plots. For most of the screen captures in this text, we will use a Pen Size of 2. There is no way to change the default Pen Size, so you will need to make the changes for each graph you create.

2.F.3. Changing Titles and Axis Labels

Each graph on a page has a title, a y-axis label (called the left axis label), and a x-axis label (called the bottom axis label):

Once again, we can obtain the properties dialog box for each graph, and change the items for each graph separately. We will work with the page with a single graph. Click the **RIGHT** mouse button on an empty space on the graph. A menu will appear:

Select **Properties**:

The **General** tab allows us to change the **Title** of the selected graph. If the **General** tab is not selected, click on the tab to select it:

In my dialog box, the text for the **Title** is selected. To change the title, all we need to do is type the new title. Type the text **Three Resistor Circuit - Voltages**:

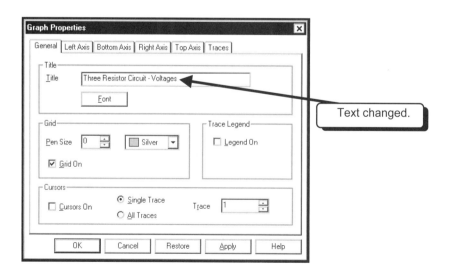

To change the y-axis settings, select the **Left Axis** tab:

EWB Multisim   The Postprocessor and the Grapher   89

This dialog box can be used to change all of the properties for the y-axis. You should experiment with the settings to see their effects. In my dialog box, the text for the **Label** is selected. To change the **Label**, all we need to do is type the new **Label**. Type the text `Voltage (Volts)`:

To change the x-axis settings, select the **Bottom Axis** tab:

This dialog box can be used to change all of the properties for the x-axis. You should experiment with the settings to see their effects. In my dialog box, the text for the **Label** is selected. Once again, all we need to do is type the new **Label**. Type the text `Time (Seconds)`:

To make the changes take effect, you can click on the **Apply** button or the **OK** button. I will click on the **OK** button. This will apply the changes and close the dialog box:

2.F.4. Adding a Second Y-Axis

The Grapher has the ability to use two y-axes on a single graph. This is a useful feature if you are trying to plot two traces with very different magnitudes on the same graph. An example would be plotting a current and a voltage. For many circuits, the voltage will be in volts while the current may be in milli- or micro-amperes. When plotted on the same y-axis scale, the current trace will look like a straight line through zero because the current is so small compared to the voltage. To fix this problem, we can use different axes for the voltage and current.

To illustrate the procedure, we will create a new Postprocessor page that plots the source voltage and source current for the circuit we simulated at the beginning of this chapter. Follow the procedures of section 2.A and use the Postprocessor to create a new page with a single graph and plot the source voltage (vin) and the source current (vvx#branch) on the same graph:

We need to obtain the Graph Properties dialog box to make the necessary changes. Click the **RIGHT** mouse button on an empty area of the graph to obtain the menu:

Select **Properties**:

On my dialog box, the **Traces** tab is selected. This is the portion of the dialog box we need to work with. Note that **Trace 1** is the input voltage and it uses the **Left Axis** as its scale. We want this trace to use the left axis, so we will not change this setting. However, we will change the **Label** and **Pen Size**:

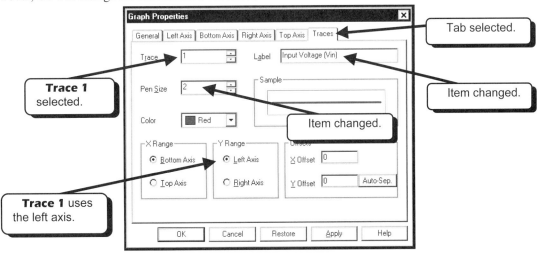

We would like trace 2 to use the right axis, which is our second y-axis. Click the **LEFT** mouse button as shown to place the cursor in the **Trace** field:

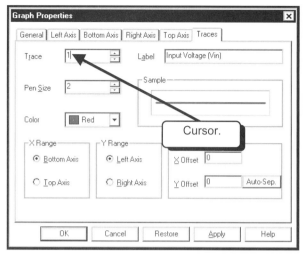

With the cursor as shown we can use the arrow keys on the keyboard to change the trace. Press the up arrow key on the keyboard to select **Trace 2**:

We see that Trace 2 also uses the left axis. We want this trace to use the right axis, so click the **LEFT** mouse button on the text **Right Axis** to select the option:

 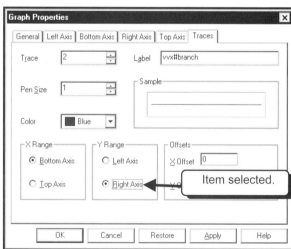

Change the **Pen Size** to **2** and the **Label** to `Source Current`:

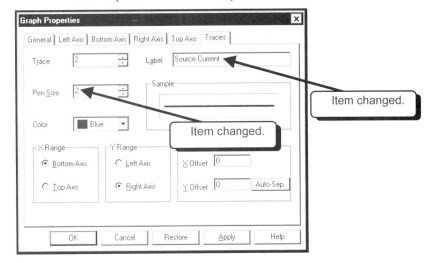

The second trace now uses the right axis. We must now look at the properties of that axis, so select the **Right Axis** tab:

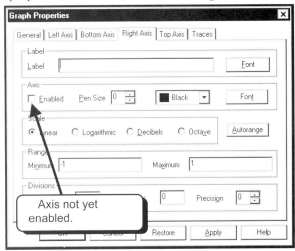

Nothing is specified for this axis. First, we must enable the axis. Click the *LEFT* mouse button on the text **Enabled** to select the option:

We note that the range is set for –1 to 1. This is not correct for the current we are plotting. We can either enter values for the range or have Multisim pick the range. Click on the **Autorange** button to have Multisim select the **Range** and **Divisions** settings.

The range goes from –1 mA up to 1 mA, and values are now specified for the tick marks. If the values are not adequate, we can change them later.

It is not obvious, but after we clicked on the **Autorange** button, the Right Axis is no longer enabled. Enable the axis again:

 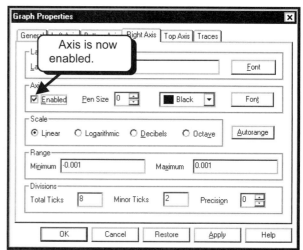

The last thing we need to do is to add a **Label** for this axis. Fill in the **Label** field as `Current (Amps)`:

To view the results of our changes, click on the **OK** button:

You can turn on the legend and the grid as shown in section 2.F.1:

2.G. Using the Cursors

The cursors allow us to obtain numerical values from a plot. We can use the cursors on a page with a single graph with a single trace, a page with a single graph with multiple traces, or a page with multiple graphs that have multiple traces. First, we will use the graph we created earlier:

There are two ways to turn on the cursors: using the ***RIGHT*** mouse button or using the Graph Properties dialog box. We will show both here. We will first use the ***RIGHT*** mouse button. Click the ***RIGHT*** mouse button on an empty area of the graph:

Select **Toggle Cursors**:

Two cursors are displayed on the page. Cursor 1 is shown in red and cursor 2 is shown in blue. By default, only information about trace 1 is displayed on the cursor information dialog box. We will look closer at this dialog box:

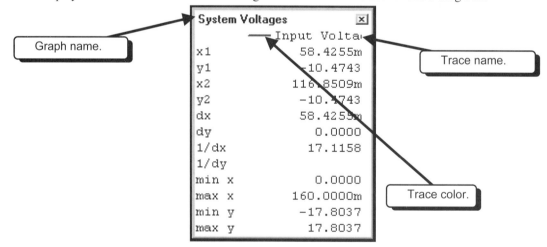

The cursor information is for the graph named **System Voltages**. Since this page contains a single graph, this information is obvious. If the page had more than one graph, this would be important information. We also see that the numerical information is displayed only for a single trace. In my case the color of the trace on the graph is purple and the name of the trace is **Input Voltage**. The remainder of the information is numerical information:

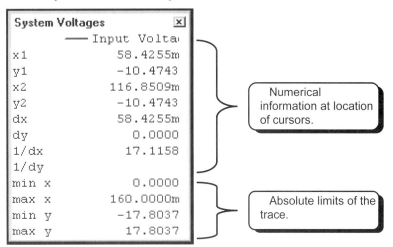

The top portion gives information about the trace at the location of the cursors. **x1** and **y1** refer to cursor 1 and **x2** and **y2** refer to cursor 2. The bottom portion gives the absolute limits for the entire trace.

If we want to display information about other traces on the graph, we need to use the Graph Properties dialog box. Click the **RIGHT** mouse button on an empty area of the graph:

Select **Properties**:

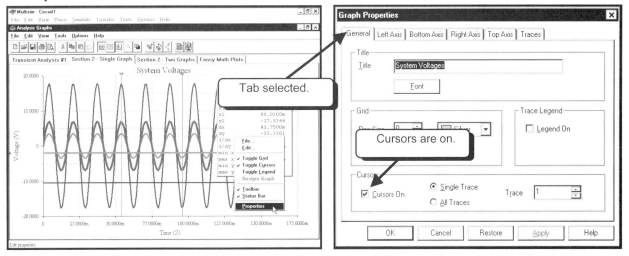

My dialog box opens with the **General** tab selected. If yours does not, click on the **General** tab to select it. Also note that the box indicates that the cursors are already on. This is because we turned on the cursors using the **RIGHT** mouse button. We can also use the **Cursors On** option to turn the cursors on or off.

This dialog box allows us to change the trace being displayed in the cursor information dialog box:

The cursor information dialog box can be set to display information for a single trace or all traces. Also, if we display information for only a single trace, we can select which trace to display. The options are currently set to display cursor information only for **Trace 1**, which happens to be the input voltage trace. Our graph has three traces, and we are interested in the numerical values for all of the traces. Click the *LEFT* mouse button on the text **All Traces** to select the option:

Click on the **OK** button to accept the change. When you return to the Grapher, the cursor information dialog box will display information for all traces on the graph:

The dialog box becomes quite large, because it displays information for every trace on the graph.

Now that we know how to display the cursor information, we need to know how to move the cursors. To move a cursor, drag the triangle at the top of the cursor. For example, to move cursor 1 (red), click and **HOLD** the *LEFT* mouse button on the red triangle as shown:

While continuing to hold down the ***LEFT*** mouse button, move the mouse. The cursor will move with the mouse and the numbers on the cursor dialog box will change. When you have found a suitable location, release the mouse button. To move cursor 2, click and **HOLD** the ***LEFT*** mouse button on the blue triangle and the move the mouse.

For our last example of using the cursors, we will use the plot with two graphs we created earlier:

Each graph on the page has separate cursors. There is a set of cursors for each graph, and the cursors are moved independently, so one graph can display cursor information for all traces on its graph while the cursors on the other graph can display information for a single trace. Thus, we treat the cursors separately for each graph. To display the cursors for the top graph, click the ***LEFT*** mouse button on an empty space in the top graph:

Select **Toggle Cursors**:

Cursors appear on the top graph, and the cursor information dialog box applies to the top graph.

To display the cursors for the bottom graph, click the **LEFT** mouse button on an empty space in the bottom graph:

Select **Toggle Cursors**:

A second cursor information dialog box appears. This one displays information for the bottom graph. You can drag these dialog boxes around the screen to position them better:

Notice that the names in the cursor information dialog boxes allow you to identify which graph they belong to:

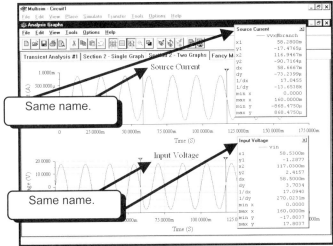

We use the same dragging method for moving the cursors. The cursors on each graph are not required to have the same time coordinate:

We see that if we have a page with multiple graphs, or a page with a single graph, the cursors go with the graph and not the page. For a page with several graphs, the cursors only apply to a single graph, and the cursors on one graph are independent of the cursors for another graph.

2.H. Zooming In and Out

There are two methods for zooming in on a trace in the Grapher. One method uses the mouse and the second method uses the Graph Properties dialog box. If you have a page with a single graph, the easiest way to zoom is to use the mouse. If you have a page with several graphs, the best way to zoom is to use the Graph Properties dialog box.

We will start with the single graph shown below:

We will first zoom in using the mouse. In this method, we will draw a rectangle with the mouse and Multisim will zoom in to the area of the rectangle. Place the mouse pointer where you would like the upper left corner of the zoom rectangle to be:

Click and **HOLD** down the *LEFT* mouse button. While continuing to hold down the *LEFT* mouse button, move the mouse. A dashed rectangle will be drawn:

Draw the rectangle so that it encloses the area in which you want to zoom:

When you release the mouse button, Multisim will zoom in on the rectangle:

We can repeat the procedure to zoom in further:

To go back to the original view, select **View** and then **Restore Graph** from the Grapher menus:

The second method we use for zooming is to change the x-axis and y-axis scales. This is tedious, but it allows us to zoom in to a precise location of the graph. Select **Edit** and then **Properties** from the Grapher menus to obtain the Graph Properties dialog box:

We will change the scale of the x-axis (**Bottom Axis**) first, so select the **Bottom Axis** tab:

We would like to zoom in on the second cycle of the waveform, so change the **Range** of the x-axis to **0.016** to **0.032**:

Next, we want to change the range of the y-axis (**Left Axis**), so select the **Left Axis** tab and change the **Range** to **-7.5** to **7.5**:

To zoom in, click on the **OK** button:

This second method may seem a bit cumbersome when compared to using the mouse. However, we must use this method when we have a page with several graphs on it. For our second demonstration, we will use the page with two graphs:

Next, use the mouse to zoom in on a portion of the top graph:

We see that Multisim only modifies the top graph. For a plot of this type we usually want the x-axes to be the same for both graphs. To accomplish this, the only thing we can do is use the Graph Properties dialog box for both graphs and set the bottom axis range to be the same for both plots. Before continuing, select **View** and then **Restore Graph** to start with the original graph.

Next, click the ***RIGHT*** mouse button on an empty space in the top graph and then select **Properties** from the menu:

Select the **Bottom Axis** tab and change the range to 0.016 to 0.032:

Click on the **OK** button to apply the changes to the top graph:

We must now repeat the procedure for the bottom graph. Click the ***RIGHT*** mouse button on an empty space in the bottom graph and then select **Properties** from the menu:

Select the **Bottom Axis** tab and change the range to 0.016 to 0.032:

Click on the **OK** button to apply the changes to the bottom graph:

We see that the x-axis scale of both graphs is the same. However, the x-axes of the two graphs do not quite line up because the text on the left side of each graph takes different amounts of space.

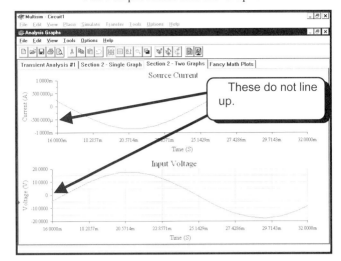

We can overcome this problem by changing the precision of the left axis on both graphs. Obtain the **Graph Properties** for the top graph and change the precision of the **Left Axis** of the top graph to 0:

Click on the **OK** button and then obtain the Graph Properties dialog box for the bottom graph. Change the precision of the **Left Axis** of the bottom graph to 2:

Click on the **OK** button to view the changes. The axes appear to line up. They are not aligned exactly, but they are close:

2.I. Saving and Opening Pages in the Grapher

The Grapher allows you to save and load graphs. This procedure saves all of the graphs currently displayed by the Grapher and includes all of the modifications we have made to the graphs. This method does not save any circuit information or Postprocessor information. Only the graphs are saved. Saving graphs does not save any information on how the graphs were created. This method just provides a quick way of saving all of the work we did in customizing the graphs.

To save the graphs, select **File** and then **Save** from the Grapher menus:

The Grapher file will be saved in the same directory as the circuit file and Postprocessor file unless you change the directory. Enter a name for the file:

Multisim will automatically add the **.gra** extension. Click on the **Save** button to save the graphs.

Close and then restart Multisim. You will be presented with an empty schematic window:

EWB Multisim   The Postprocessor and the Grapher   111

Type **CTRL-G** to open the Grapher. You will be presented with an empty Grapher window. Maximize the Grapher window so that it occupies the entire Multisim window:

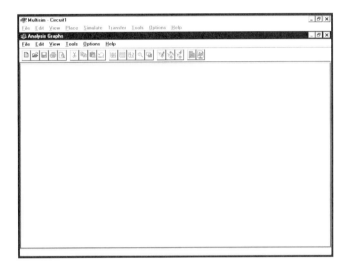

Select **File** and then **Open** from the Grapher menus and select the file we just created:

When you click on the **Open** button, the pages we saved earlier will be displayed:

When we display the other pages, we see that they contain the changes we made using the Grapher facilities:

We cannot add any more traces to the plots, nor add any more pages to the Grapher with the file we just opened. If you want to add more traces for the circuit we simulated, you will need to open the circuit file, run the simulation, and then use the Postprocessor to create more graphs.

2.J. Problems

Problem 2-1: When you set up a SPICE simulation such as a Transient Analysis, what parameters must you specify so that you can view results using the Postprocessor and Grapher?

Problem 2-2: What does the **Plot during simulation** button do?

Problem 2-3: What is the difference between an output variable such as vin or vout and an output variable such as vvx#branch?

Problem 2-4: What is the difference between the Grapher and the Postprocessor?

Problem 2-5: What is the difference between a page, a graph, and a trace?

Problem 2-6: Choose five functions available in the Postprocessor and describe their mathematical function. Example functions are abs, mean, and real.

Problem 2-7: What is the procedure for creating a new page with a single graph on it?

Problem 2-8: What is the procedure for adding a new trace to a page that has several graphs?

Problem 2-9: What is the difference between the Delete Page, Delete Diagram, and Delete Trace buttons?

Problem 2-10: What is the difference between a graph and a chart?

Problem 2-11: In the Grapher window below, how can you tell which graph is selected? This indicator was not covered in the manual.

Problem 2-12: Specify the procedure for deleting a page in the Grapher window.

Problem 2-13: For this problem, use the circuit simulated in Part 2. With the Postprocessor, create a new page that has three graphs. The top graph plots the instantaneous power absorbed by Rx using the equation V^2/R, the middle graph plots the instantaneous power absorbed by Rx using the equation I^2R, and the bottom graph plots the instantaneous power absorbed by Rx using the equation $V \times I$.

Problem 2-14: For this problem, use the circuit simulated in Part 2. With the Postprocessor, create a new page that has one graph. Plot two traces. The first trace is the instantaneous power supplied by the source. The second trace is the instantaneous power absorbed by Rx plus the instantaneous power absorbed by Ry plus the instantaneous power absorbed by R1 plus the instantaneous power absorbed by R2. Show that the two traces are the same.

Problem 2-15: For this problem, use the circuit simulated in Part 2. With the Postprocessor, create a new page that has one graph. Plot two traces. The first trace is the average power supplied by the source. The second trace is the average power absorbed by Rx plus the average power absorbed by Ry plus the average power absorbed by R1 plus the average power absorbed by R2. Show that the two traces are the same.

Problem 2-16: What is the difference between saving information in Multisim, the Postprocessor, and the Grapher?

Problem 2-17: What are the three file types created by Multisim?

Problem 2-18: Under what circumstances would you add a second y-axis to a graph?

Problem 2-19: Load the pages saved using techniques in section 2.D. On the page with three graphs, modify the page so that there is a grid on all three graphs, and a legend on the top two graphs but not the bottom graph. Turn in a printout of your solution.

Problem 2-20: In the Grapher page that contains three graphs, change the title of the middle plot to "My Middle Graph," change the y-axis label of the top graph to "Strawberries," and change the x-axis label to "Time – Warp." Turn in a printout of your solution.

Problem 2-21: What is the difference between selecting the **OK** button and selecting the **Apply** button in the Graph Properties dialog box of the Grapher?

For Problems 2-22 through 2-26, start with the graph below:

Problem 2-22: Modify the grid of a graph so that the pen size of the grid is 2 and the color of the grid is red. Turn in a printout of your solution.

Problem 2-23: Modify the left axis of a graph so that the pen size of the axis is 2 and the color is blue. Turn in a printout of your solution.

Problem 2-24: Modify the left axis of a graph so that the total number of ticks is 16 and the number of minor ticks is 4. Turn in a printout of your solution.

Problem 2-25: Modify the bottom axis of a graph so that the pen size of the axis is 3 and the color is green. Turn in a printout of your solution.

Problem 2-26: Modify the bottom axis of a graph so that the total number of ticks is 14 and the number of minor ticks is 2. Turn in a printout of your solution.

Problem 2-27: What is the maximum number of graphs you can put on a single Postprocessor page? What is the maximum number of graphs you can put on a single Postprocessor page and still have a printout that looks useful? Turn in printouts supporting your answer.

Problem 2-28: Specify three methods for showing or hiding cursors on a Grapher page.

Problem 2-29: Specify three methods for showing or hiding the grid on a Grapher page.

Problem 2-30: Specify three methods for showing or hiding the legend on a Grapher page.

Problem 2-31: Specify two ways to obtain the Graph Properties dialog box.

Problem 2-32: Describe the two methods for zooming in the Grapher.

Problem 2-33: List three ways of saving files and information in Multisim.

Problem 2-34: What are the advantages and disadvantages of using the two different zoom methods covered in section 2.H?

Problem 2-35: Specify two ways of deleting pages in the Grapher.

PART 3
DC Measurements

In this chapter we will find the DC values of currents and voltages in a circuit. For the analyses we show here, the DC voltages will not change. In Chapter 4 we will look at the DC sweep analysis where you can sweep a DC source through several values. The simulations are similar in that the values displayed are DC values. With a DC sweep, however, you can see how the DC values of a circuit change as a DC source in the circuit is changed. For example, we could plot how the bias current of a transistor changes for different values of the DC supply. In contrast, in this chapter we can only display the collector current for a single value of the supply. To find the bias current at a different value of the DC supply we would have to change the value of the supply and then rerun the simulation. We will show several ways to view the results, including using the current indicators, the Multimeter, and running a DC Operating Point Analysis.

3.A. Resistive Circuits

We will perform simulations of the circuit below:

3.A.1. Measurements with Indicators

We would first like to find the current supplied by **V1** and the voltage across resistors **R2** and **Rz**. To measure current and voltage, we use the voltage and current indicators in the Indicator toolbar. Click the *LEFT* mouse button on the **Indicators** button:

To place a voltage monitor, click the on voltage indicator button. You will be presented with a few different styles of voltage indicators:

All indicators perform the same function. The terminals are arranged differently to make them more convenient to place in your circuit. Choose an indicator, place two in your circuit, and wire them as shown. Note that voltage is measured between two nodes in a circuit:

To place a current indicator, click on the current indicator button , which is also located in the Indicators button bar. You will be presented with a few different styles of current indicators:

A current indicator must be placed in series with the element through which you want to measure the current. Choose a current meter and add it as shown:

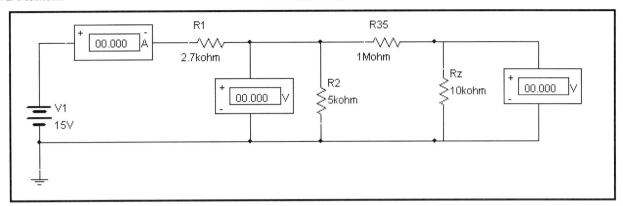

It is important to understand what the plus and minus signs mean in the current indicator. If the current enters the positive terminal and leaves the negative terminal, the monitor will read a positive number. That is, if the current flows in the direction shown below, the current indicator will read a positive number:

In the circuit shown above, the current indicator will measure the current supplied by V1 or the current flowing through R1.

Before we simulate the circuit, we will look at the current and voltage indicators a little more closely. First, double-click the **LEFT** mouse button on the current indicator. This will open the properties dialog box for the indicator:

Select the **Value** tab if it is not already selected.

There are two important parameters for this part. First, the **Mode** is set to **DC**. This means that the indicator will display the **DC** or average value of the current it measures. Since all of our sources in this circuit are DC, this is the correct setting for this application. The other parameter is the series resistance of the current indicator. All real current meters are placed in series with the element through which they measure current, and all current meters have a small series resistance. Since the indicator is placed in series with another circuit element, the resistance of the indicator will affect the circuit because its resistance adds to the overall resistance of the circuit. To reduce the effects of the indicator a small series resistance is required. To provide an accurate model, Multisim allows you to specify the value of this series resistance.

There are two ways to choose a value for this resistance. If you are comparing this simulation to a real circuit where you measure the current with a real ammeter, you should set the resistance in the simulation to the resistance of your real ammeter. The manuals that accompany your meter should have a specification of its resistance. This will accurately model your ammeter in the circuit. If the ammeter has an effect on the circuit, your simulation will include this effect. The second method of choosing the current indicator resistance is to choose it small so that the series resistance does not affect the circuit. The default value of this series resistance is 1 nΩ (1e-009 ohm). This resistance is small compared to all other resistances in the circuit, so the current indicator resistance will have a negligible effect on the circuit. We are not comparing this circuit to a real-world measurement, so we will use the default value of 1 nΩ. Click on the **OK** button to close the dialog box.

Next, we will look at the properties of the voltage indicator. Double-click the *LEFT* mouse button on the voltage indicator. This will open the properties dialog box for the indicator:

Select the **Value** tab if it is not already selected.

As with the current indicator, you can set the voltage indicator to measure AC or DC voltages. We will leave the **Mode** set at **DC**. The other parameter is the parallel resistance of the voltage indicator. All real voltage meters are placed in parallel with the element through which they measure current. In order to measure the voltage, the meter must draw a small current from the nodes to which it is connected. The current the meter draws is modeled as a resistance between the meter terminals. Since the voltage indicator is connected between two nodes, the resistance of the indicator will draw current and affect the circuit. To reduce the effects of the voltage indicator, a large resistance is required. To provide an accurate model, Multisim allows you to specify the value of this resistance. The default is 1 MΩ. If the current the meter draws is small compared to other elements in the circuit, the meter will not affect the circuit very much. If the meter resistance is on the order of other elements in the circuit, the act of making a measurement may cause a large change in the circuit, and the measurement will be inaccurate. For now, we will assume that 1 MΩ is large enough and we will not change the value. We will check later to see if this is indeed the case. Click on the **OK** button to close the dialog box.

We are now ready to simulate the circuit. Select **Simulate** and then **Run** from the Multisim menus, press the **F5** key, or click on the simulate button :

When you simulate the circuit, the indicators will display their measured values:

EWB Multisim   DC Measurements   119

Next, we will check to see if the resistance of the voltage indicators had an effect on the readings. The first thing we need to do is to stop the simulation. Select **Simulate** and then **Run** from the Multisim menus, press the **F5** key, or click on the simulate button [icon] to toggle the simulation off. You can tell that the simulation has stopped running because the time indicator at the bottom of the screen should stop counting:

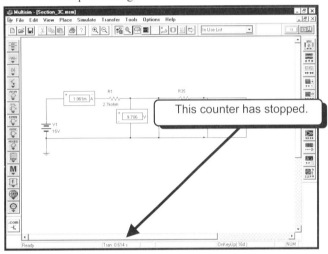

Double-click the **LEFT** mouse button on a voltage indicator and select the **Value** tab. Change the resistance to 1 GΩ as shown below:

Note that 1 GΩ equals $1\times10^9 \Omega$. This should be large enough compared to other elements in the circuit so that it draws a negligible amount of current. Click on the **OK** button to close the dialog box. Modify the other voltage indicator in the circuit and change its resistance to 1 GΩ as well. When you are done, the circuit will look unchanged:

Note that the numbers displayed by the indicators will not change until we run the simulation. Select **Simulate** and then **Run** from the Multisim menus, press the **F5** key, or click on the simulate button to start the simulation. In a few moments, the indicators may change their display:

We see that the numbers on all of the indicators have changed slightly. The indicators now read 1.956 mA, 9.723 V, and 0.096 V. These values are all slightly different from the original readings of 1.961 mA, 9.706 V, and 0.095 V. These changes may or may not be significant depending on your application. The important thing to note is that whenever you measure a quantity, the act of making that measurement may affect the quantity you are measuring. Multisim accurately models these effects by giving a current indicator a series resistance and a voltage indicator a parallel resistance. The theoretical value of the current is 1.954 mA, so we see that even the high-resistance meters we created do affect the circuit behavior slightly.

Note that an ideal current indicator would have no series resistance, and an ideal voltage indicator would have an infinite parallel resistance. Be advised that if you make the resistance of a current indicator too small, it will not read the correct current. The screen capture below shows the result of setting the current indicator resistance to 1 pΩ, or 10^{-12} Ω:

The current is improperly displayed as 3.553 mA. These examples show us that you should be aware of the effects of using an indicator, and that you cannot adjust its resistance to make it ideal either. Note that a series resistance of 1 nΩ is close to ideal for the current monitor while a parallel resistance of 1 MΩ for a voltage indicator is on the small side. When you use a voltage monitor, you should consider increasing its resistance to 1 GΩ so that it has less effect on your circuit.

EXERCISE 3-1: Find the unknown node voltages and the current through R1.

SOLUTION: Add voltage and current indicators as shown below and run the simulation.

EXERCISE 3-2: Find the DC node voltages for the circuit below and the DC current through R1:

HINT: A DC current source is located in the Sources button bar:

SOLUTION: Add indicators as shown and run the simulation:

3.A.2. Measurements with the Multimeter

You can also measure voltage and current using a model of a Multimeter. We will start with the original circuit, which is repeated below:

To place a Multimeter, click the **LEFT** mouse button on the Multimeter button as shown above. A meter will become attached to your mouse pointer.

Place the meter as you would any part. We will measure a voltage and a current, so we will need two meters in our circuit:

We will measure the current through **R1** using meter **XMM2** and the voltage across **R2** using **XMM1**. When you wire up a current meter, you must place the meter in series with the element through which you want to measure the current. When you measure a voltage, you place the meter in parallel with the element across which you want to measure the voltage. Wire the meters as shown:

It is important to understand what the plus and minus signs mean when you measure current with a Multimeter. If the current enters the positive terminal and leaves the negative terminal, the monitor will display a positive number. That is, if the current flows in the direction shown below, the current indicator will display a positive number:

To view the display and function of a meter, double-click the **LEFT** mouse button on the meter. A window will open displaying the meter:

We see that we can use the meter to measure current, voltage, resistance, and decibels. We can also select between AC and DC measurements. On my monochrome screen captures it is not easy to see which mode is selected. If you click on the **Set** button on the meter, you will see the model parameters for this meter:

Here we see model components similar to those for the voltage and current indicators on page 117. When the meter is set to measure current, it has a resistance of 1 nΩ. When the meter is set to read voltage, the meter looks like a large resistance of 1 GΩ. When the meter is used to measure resistance, it passes a 10 nA current through the device it is connected to, and it reads the voltage across that element. These resistances make the meters close to ideal, and we will not change them. It is important to note that if a series resistance of 1 nΩ or a parallel resistance of 1 GΩ is significant in your circuit, you will need to account for these resistances in your circuit analysis. Click on the **OK** button to close the **Multimeter Settings** dialog box.

We can view both meters at the same time, so double-click the **LEFT** mouse button on the other Multimeter to open its window. You should have both Multimeters opened as shown below:

EWB Multisim   DC Measurements   125

Select the left meter to measure current by clicking on the **A** button [A]. Select the right meter to measure voltage by clicking on the **V** button [V].

We can now start the simulation. Select **Simulate** and then **Run** from the Multisim menus, press the **F5** key, or click on the simulate button [icon] to start the simulation. The meters will display their measured values:

Note that you can press the meter buttons while the simulation is running. I will change **XMM2** to read voltage:

After a few moments the meters change their values, but we notice that the value displayed by **XMM1** also changes. This is because when we changed **XMM2** to read voltage, the model of the meter changes to a 1 GΩ resistor. This large resistance dominates the circuit and causes all of the readings to change. This is obviously not the correct way to use this meter, so I will change **XMM2** back to reading current.

We can also use these meters to read AC or DC waveforms. Switch the meters to AC by clicking on the AC button ⌐∿⌐ as shown below:

We see that both meters change their display after a few moments and read 0. This is because there are no AC sources in the circuit, so the measured items have no AC component and therefore read 0. You can close the meter displays by clicking on the ⌐X⌐ in the upper right corner of the Multimeter windows:

EXERCISE 3-3: Find the unknown node voltages and the current through R1 using Multimeters.

SOLUTION:

EXERCISE 3-4: Find the DC node voltages for the circuit below and the DC current through R1 using meters.

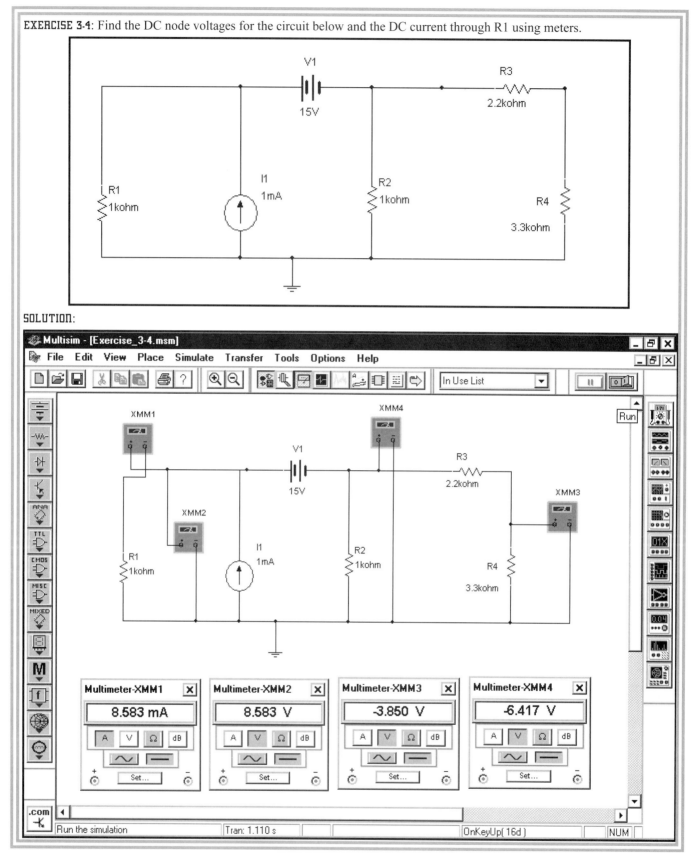

SOLUTION:

3.A.3. Using the Power Meter

Another meter we can use in the circuit of the previous section is the power meter. We will start with the circuit below:

EWB Multisim   DC Measurements   129

Double-click on the Wattmeter button in the instruments toolbar. A wattmeter will become attached to your mouse pointer:

We would like to display the power absorbed by R1, so place the meter as shown below:

To measure the power absorbed by R1, we must measure the current through the resistor and the voltage across the resistor. Measuring current with the wattmeter is the same as measuring current with a Multimeter or with a current indicator; if current enters the positive terminal and leaves the negative terminal, the meter will interpret it as a positive current. Wire the meter as shown:

As shown, the meter measures the left terminal of R1 as the positive voltage reference, and current is positive flowing left to right. Thus, the positive direction of current is entering the positive voltage terminal, and this is the definition of power, P = VI. Note that P = VI applies only if you follow the convention that the positive direction of current enters the positive voltage reference terminal.

Double-click the **LEFT** mouse button on the power meter to open its window:

This meter works for AC or DC measurements. Select **Simulate** and then **Run** from the Multisim menus, press the **F5** key, or click on the simulate button to start the simulation.

We see that R2 absorbs **10.312 mW** of power.

EXERCISE 3-5: Find the power absorbed by R3 in the circuit below:

SOLUTION: Add the power meter as shown:

EXERCISE 3-6: Find the power supplied by I1:

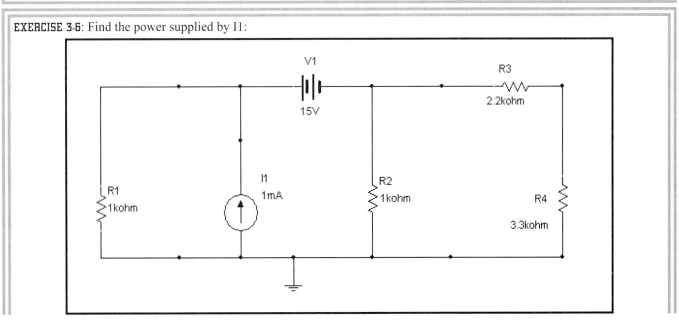

SOLUTION: Redraw the circuit to make the placement of the power meter more convenient.

3.A.4. DC Operating Point Analysis

The final way of simulating a circuit and finding the DC node voltages we will look at is the DC Operating Point Analysis. We will start with the example circuit used in previous sections:

This analysis gives us the voltages at all nodes of the circuit. However, we do not know how the nodes are numbered. We can display the node names in two ways. One way is to double-click on a wire. For example, double-click the *LEFT* mouse button on a wire connected to the middle node:

This dialog box tells us that this node is numbered 13 (your name may be different) and it also allows us to change the node name. I will change the name to **Vmid**:

Click on the **OK** button to save the change:

We now know the name of this node, but it is still not displayed on the circuit.

To display the node names on the schematic, select **Options** and then **Preferences** from the Multisim menus:

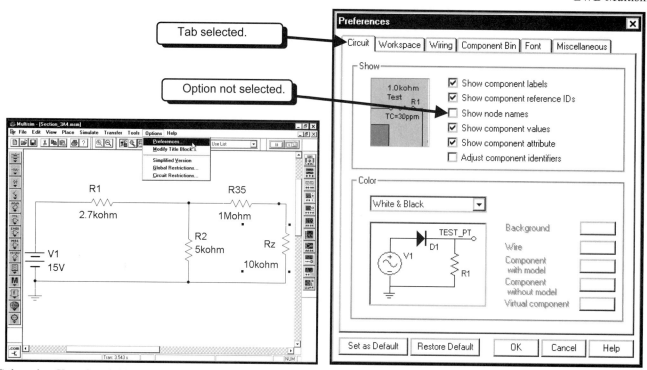

Select the **Circuit** tab if it is not already selected. Note in the dialog box above that the option to display node names is not selected. To display node names on the circuit, we need to select this option. Click the *LEFT* mouse button on the empty box ☐ next to the **Show node names** option to enable it. When the option is enabled, the box will have a checkmark ☑:

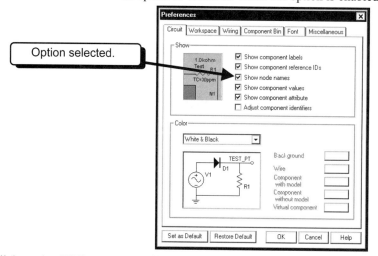

When you click on the **OK** button, you will return to the schematic and the node names will be displayed.

The node numbering of your circuit may be different. We now know the names and can set up the DC Operating Point simulation. Select **Simulate**, **Analyses**, and then **DC Operating Point** from the Multisim menus:

The only thing we need to do is to specify the nodes we are interested in. To select a node, click the **LEFT** mouse button on the node to select it. For example, click the **LEFT** mouse button on the text 2 as shown. It will become highlighted, indicating that it has been selected:

Click on the **Plot during simulation** button. The node will be added to the list of nodes whose result is displayed:

Repeat the procedure to add nodes **3** and **vmid**:

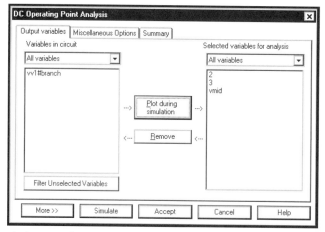

We are now ready to run the simulation, so click on the **Simulate** button. The results are displayed in a new window:

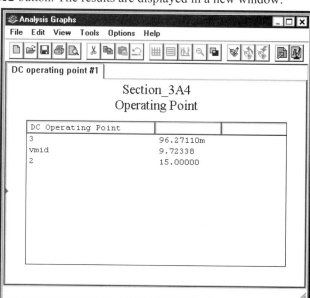

We see that the voltage at node **3** is **96.27** mV, the voltage at node **Vmid** is **9.72** volts, and the voltage at node **2** is **15** volts. These results agree with the results of the previous analyses.

3.B. Nodal Analysis with Dependent Sources

To illustrate an example with dependent sources, we will find the node voltages in the circuit below. Create the circuit below.

The new part in this circuit is the voltage-controlled voltage source. The way this element is wired, the voltage at node **Vc** is 5 times the voltage across **R1**: $V_c = 5(V_b - V_a)$. If you zoom in on the voltage-controlled voltage source, you will see the graphic shown in Figure 3-1. The plus (+) and minus (−) terminals on the voltage-sensing element (the left element) are open circuited. These connections draw no current and only sense the voltage of the nodes to which they are connected. The right half of the graphic contains the dependent source. The voltage of this source is the gain 5V/V times the voltage at the sensing nodes.

Figure 3-1: Voltage-dependent voltage source.

We can use any of the methods presented in section 3.A to measure the node voltages. We will use a number of voltage indicators to measure various voltages. You can use the DC Operating Point analysis or the meters to measure the voltages if you wish. My modified circuit is shown below:

When you have made the changes to your circuit, click the **LEFT** mouse button on the Run button to simulate the circuit. The indicators will be updated and display the DC node voltages:

It turns out that this circuit is fairly sensitive to the input impedance of the voltage indicators. We will change the input impedance of the indicators from the default value of 1 MΩ to 1 GΩ. First, we must stop the simulation. Click the *LEFT* mouse button on the Run button to turn the simulation off. Next, double-click the *LEFT* mouse button on one of the voltage indicators as shown below:

We see that the default value of the parallel resistance is 1 MΩ. To compare the simulated results to calculated results, we need to increase the parallel resistance. Change this value to 1 GΩ or higher and then click on the **OK** button. Increase the resistance of all voltage indicators in your circuit to 1 GΩ or higher. Click on the Run button to simulate the circuit and update the displays of the indicators:

We see that the numbers displayed now are different than before. For most of your simulations, you should increase the voltage indicator parallel resistance to 1 GΩ or higher, or the voltage indicators will affect the operation of the circuit.

EXERCISE 3-7: Find the DC node voltages for the circuit below:

HINT: I1 is a current-controlled current source. Note that node 3 is connected to ground with a wire. Thus, the voltage of node 3 should be zero volts. Node 3 is necessary because it joins the lower terminal of R2 to the current-sensing terminal of I1.

SOLUTION: Use meters or voltage indicators to measure the voltages of the unknown nodes. The voltage at node 1 is known and is equal to V1.

EXERCISE 3-8: Find the DC node voltages for the circuit below:

HINT: **I1** is a voltage-controlled current source. The current through **G1** is 0.5 times the voltage at node 1 minus the voltage at node 4.

SOLUTION: Add voltage indicators to the circuit and change their parallel resistance to 1 GΩ.

3.C. Diode DC Current and Voltage

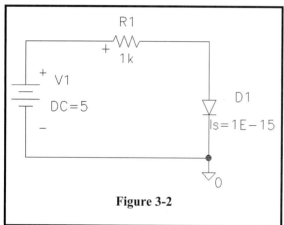

Figure 3-2

We will now use Multisim to find the diode current and voltage in the circuit of Figure 3-2. The diode current is given as $I_D = I_S[\exp(V_D/\eta V_T) - 1]$. I_S is the diode saturation current and is 10^{-15} amp for this example. V_T is the thermal voltage and is equal to 25.8 mV at room temperature. η is the emission coefficient for the diode and its default value is 1. Multisim automatically runs all simulations at room temperature by default.

When you use a diode in a circuit, you will have to specify a model for the diode. In our case, the model will tell Multisim the value of I_S for our diode. The libraries have a number of predefined models that are usually used in a classroom environment. However, the model for this diode is not in our libraries, so we will have to define a new model for it. The part for the diode you should use is **DIODE_VIRTUAL**. Draw the circuit below:

The **DIODE_VIRTUAL** model is located in the Diodes toolbar as shown below. To display the Diodes toolbar, click on the diode button as shown:

DC Measurements

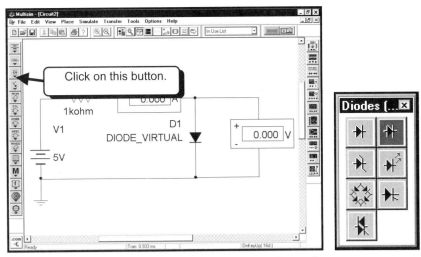

The **DIODE_VIRTUAL** button is the one in the upper right corner of the button bar:

Parts that are referred to as "virtual" by Multisim are ideal parts, or parts that are easy to specify by a value or model. The virtual diode implements the basic diode equation $I_D = I_S[\exp(V_D/\eta V_T) - 1]$ with η equal to 1. All other model parameters are set to their default.

After placing the diode in your circuit, we need to change the model parameter I_S. Double-click on the diode graphic:

Click on the **Edit Model** button:

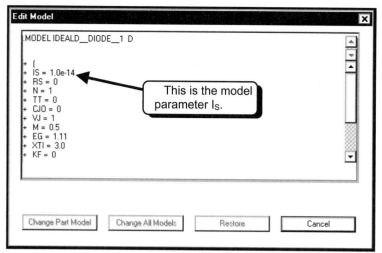

We see that the default value for the saturation current is Is = 10^{-14} amp. Change the value to 1.0e-15:

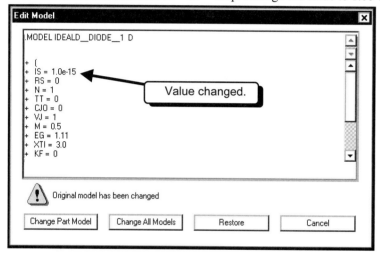

Click on the **Change Part Model** button and then click on the **OK** button. You will return to the schematic.

The schematic displays an asterisk next to the part name to indicate that we have changed the model. Run the simulation by clicking on the Run button [icon]. The diode current and voltage will be displayed:

We see that the diode current is 4.248 mA, and the diode voltage is 0.752 volts. You may need to change the resistance of your indicators to match the numbers shown.

EXERCISE 3-9: Find the diode current and voltage in the circuit below:

SOLUTION: Use the same diode parameters as in the previous example. Add indicators and then simulate the circuit.

The diode voltage is 0.752 V, and the diode current is 4.248 mA. The diode voltage and current of this circuit are the same as the diode voltage and current of the previous example. This result should be expected since the Thevenin equivalent of V1, R1, R2, and R3 in this exercise is exactly the same as the circuit of the previous example.

3.C.1. Changing the Temperature of the Simulation

In the last simulation we found the diode voltage and current at the default temperature of 27°C. Suppose we want to simulate the circuit at a different temperature? This can easily be done by changing the default instrument settings. From the Multisim menus select **Simulate** and then **Default Instrument Settings**:

Select the **Miscellaneous Options** tab:

This dialog box lists most of the options you can change. Scroll down the list until you see the **temp** option:

Click the *LEFT* mouse button on the text **temp** to select the option:

To modify the value of this option, click the **LEFT** mouse button on the square □ next to the text **Use this option**. The box should become filled with a checkmark ☑, indicating that the option is selected:

When you select the option, a value appears in the **Option Value** text field. For temperature, the units of this option are degrees centigrade. By default, all simulations are run at 27°C unless you specify a different temperature. For this run we will find the diode voltage and current at 125°C. Change the **Option Value** as shown:

Click on the **Accept** button to accept the value and return to the schematic. Run the simulation by clicking on the Run button . The diode current and voltage will be displayed for a temperature of 125°C:

3.D. Finding the Thevenin and Norton Equivalents of a Circuit

Multisim can be used to easily calculate the Norton and Thevenin equivalents of a circuit. The method we will use is the same as if we were going to find the equivalent circuits in the lab. We will make two measurements, the open circuit voltage and the short circuit current. The Thevenin resistance is then the open circuit voltage divided by the short circuit current. This will require us to create two circuits, one to find the open circuit voltage, and the second to find the short circuit current. In this example, we will find the Norton and Thevenin equivalent circuits for a DC circuit. This same procedure can be used to find the equivalent circuits of an AC circuit (a circuit with capacitors or inductors). However, instead of finding the open circuit voltage and short circuit current using DC sources, we would use AC sources.

For this example, we will find the Thevenin and Norton equivalent circuits for the circuit attached to the diode in **EXERCISE 3-9**. The circuit is repeated below:

This circuit is difficult because it contains a nonlinear element (the diode) and a complex linear circuit. If we could replace V1, R1, R2, and R3 by a simpler circuit, the analysis of the nonlinear element would be much easier. To simplify the analysis of the diode, we will find the Thevenin and Norton equivalent circuits of the circuit connected to the diode; that is, we will find the Thevenin and Norton equivalents of the circuit below:

We will convert this circuit into the Thevenin equivalent :

Once we find numerical values for Vth and Rth, the entire circuit of **EXERCISE 3-9** reduces to:

For determining the diode voltage and current, this circuit is much easier to work with than the original. This example is concerned with finding the numerical values of the equivalent circuit. The analysis of the circuit above was covered in Section 3.C. We will now find the Thevenin and Norton equivalent circuits of the circuit shown below:

We will first find the open circuit voltage. This is just the voltage across the two terminals in the circuit shown above. We will add a ground to the circuit and a voltage indicator:

The lower terminal is now at ground potential, or zero volts. Before we run the simulation, we need to change the attributes of the voltage indicator. Double-click the **LEFT** mouse button on the voltage indicator graphic to edit its properties:

Select the **Value** tab as shown above and then change the resistance to 1 TOhm (10^{12} Ω).

This resistance is large enough to be considered an open circuit compared to the other resistors in the circuit. Click on the **OK** button when you have changed the value. Run the simulation by pressing the **F5** key:

The open circuit voltage is **5.000** volts.

Next we must find the short circuit current. All we need to do is replace the voltage indicator with a current indicator:

Current measuring devices have a very low resistance so that they have a negligible effect on the current they are measuring. The current indicator in Multisim has a series resistance of 1 nΩ, or 10^{-9} Ω. If we place a current indicator across two terminals, as shown above, we are essentially placing a 1 nΩ resistor between the two terminals. This resistance is small enough to be considered a short circuit. Thus, the current indicator shown above measures the short circuit current. All we need to do now is run the simulation:

We see that the short circuit current is **5.000** mA.

We can now find the Thevenin resistance by dividing the open circuit voltage by the short circuit current:

$$R_{th} = \frac{Voc}{Isc} = \frac{5.000 \text{ V}}{5.00 \text{ mA}} = 1000 \, \Omega$$

Our Thevenin and Norton equivalent circuits are shown below:

EXERCISE 3-10: Find the Thevenin and Norton equivalent circuits for the circuit below:

SOLUTION: Voc = 4.25 V, Isc = 8.5 mA, Rth = 500 Ω. Use the circuits below:

3.E. Transistor DC Operating Point

One of the first things you should do when you are simulating a transistor amplifier circuit is to check the transistor operating point, also referred to as the transistor DC Operating Point. If the transistor operating point is incorrect, none of the other analyses will be valid. If another analysis does not make sense, check the operating point to see if all active devices are biased correctly. When Multisim finds the operating point, it assumes that all capacitors are open circuits and that all inductors are short circuits.

For this example we will demonstrate the DC Operating Point of an NPN BJT. The same procedures could be used for a PNP BJT, a jFET, or a MOSFET. We will show two ways to view the data: using the audit trail, and displaying values in a simulation output window. We will illustrate the analysis with the circuit below:

The 2N3904 transistor is located in the **Transistors** toolbar. Click on the Transistors button as shown to display the **Transistors** toolbar:

This toolbar is broken down into the different types of transistors available. A 2N3904 is an NPN bipolar junction transistor.

The NPN transistor button is located in the upper left corner of the Transistors button bar. Select this button to see the list of available NPN transistors:

Select the **2N3904** and click on the **OK** button to place the part. When you finish drawing your circuit, you can continue with the procedure below.

We must first set up the DC Operating Point simulation. Select **Simulate**, **Analyses**, and then **DC Operating Point** from the menus:

The first method we will use to display the operating point is the audit trail. Click on the **More>>** button:

EWB Multisim   DC Measurements   153

We need to enable the option **Show all device parameters at end of simulation in the audit trail**. Click on the ☐ next to this option to enable it. The square should fill with a checkmark ☑:

Although we are not interested in any of the node voltages, we need to specify an output variable in order for the simulation to run. Click on the text **1** as shown to select the output variable:

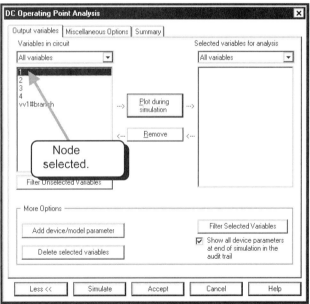

Click on the **Plot during simulation** button to add the node to the list of output variables that will be displayed during the simulation:

154   DC Measurements   EWB Multisim

Now that we have an output variable listed, we can run a simulation. Click on the **Simulate** button to run the simulation:

A new window pops up and displays the results of the simulation. The only value shown is the DC voltage at node 1. The information we seek is displayed in the audit trail, so we can close this window. Click on the ☒ in the upper right corner of the window to close it:

To view the audit trail, select **View** and then **Show Simulation Error Log/Audit Trail** from the Multisim menus:

Scroll up or down in the log until you see the information about the BJT:

We see that the collector current (**ic**) is 13.9 mA, the base current (**ib**) is 80.4 μA, **vbe** is **0.736** volts, and **Vbc** is **-6.25** volts. Vce is not directly displayed, but Vce is equal to –Vbc + Vbe, which is equal to 6.986 volts.

If we scroll down in the list, we can see the parameters for the hybrid-π small-signal model:

In the hybrid-π model we are used to resistances such as r_π, r_o, r_x and r_μ. The model parameters shown here are conductance parameters with $r_\pi = 1/$**gpi**, $r_o = 1/$**go**, $r_x = 1/$**gx**, and $r_\mu = 1/$**gmu**. There is more information displayed in the audit trail, but we will not look at it now. Click on the ☒ in the upper right corner of the window to close it:

Next, we will run the DC Operating Point Analysis, but we will display the results differently. From the Multisim menus, select **Simulate**, **Analyses**, and then **DC Operating Point** from the menus:

We would like to display the transistor currents, voltages, and some small-signal quantities found in the simulation. A problem is that these quantities are not yet displayed in this dialog box and we need to add them. Click on the **More>>** button:

Click on the **Add device/model parameter** button:

There is only one BJT device in this circuit, so this dialog box selects that device and refers to it as **qq1**. We can display the device model parameters or the device parameters. A model parameter would be something like a diode's saturation current or a transistor's current gain, β. A device parameter would be something like a transistor's collector current or base-emitter voltage. We are interested in the transistor's currents, voltages, and small-signal model parameters, and these quantities are listed as device parameters. By default this dialog box is set to display the device parameters of **qq1**, so all we need to do is select those parameters. Click the *LEFT* mouse button on the down triangle ▼ as shown:

A list of device parameters will be displayed:

Scroll down the list until you see the parameter **ic**. This is the transistor's collector current:

We see several parameters that we are interested in. Click the **LEFT** mouse button on the text **ic** to select it:

The dialog box tells us that this parameter is the collector current. Click on the **OK** button to add the parameter to the list:

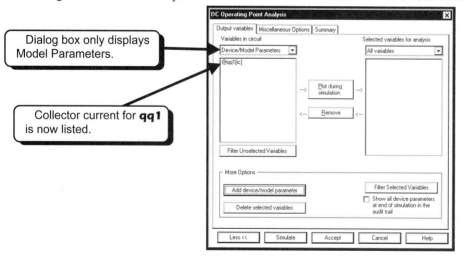

The collector current for **qq1** is now listed, but we have not yet selected it to be displayed when the simulation runs. We need to specify that we want this parameter displayed in the output. Click the *LEFT* mouse button on the text **@qq1[ic]** to select it:

Click the *LEFT* mouse button on the **Plot during simulation** button:

The bias collector current will now be displayed when the DC Operating Point Analysis runs. Repeat the procedure to display the base current, Vbe, Vbc, gpi, and gm for this device:

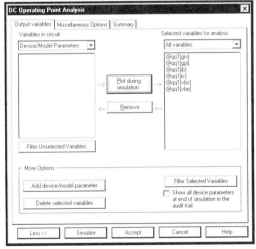

We are now ready to run the simulation. Click on the **Simulate** button. The results are displayed in the **Analysis Graphs** window:

These numbers agree with the information displayed in the audit trail.

EXERCISE 3-11: Run the DC Operating Point Analysis for the transistor circuit and display the collector current, base current, Vbc, Vbe, gpi, and gm for the BJT.

SOLUTION:

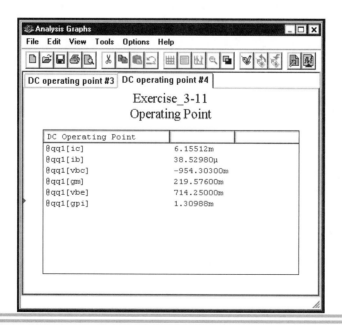

EXERCISE 3-12: Run the DC Operating Point Analysis for the circuit below and display the drain current, gate current, Vgs, Vgd, and gm for the jFET.

SOLUTION: You will need to click on the **Filter Unselected Variables** button and enable all of the options to display the required parameters. You must enable the options because the model for the 2N5486 is a subcircuit and these items are not usually displayed.

3.F. Problems

Problem 3-1: For the circuit shown below, find the unknown node voltages and the current through R3 using indicators.

Problem 3-2: For the circuit of Problem 3-1, find the unknown node voltages and the current through R3 using Multimeters.

Problem 3-3: For the circuit of Problem 3-1, find all node voltages using the SPICE DC Operating Point Analysis.

Problem 3-4: For the circuit of Problem 3-1, find the power absorbed by R6 using the power meter.

Problem 3-5: For the circuit of Problem 3-1, find the current through all resistors using the indicators.

Problem 3-6: For the circuit of Problem 3-1, find the power absorbed by R4, R5, and R6 using the power meter.

Problem 3-7: For the circuit of Problem 3-1, find the power supplied by V1 using the power meter.

Problem 3-8: For the circuit of Problem 3-1, find the unknown node voltages and the current through R2 using indicators.

Problem 3-9: For the circuit of Problem 3-1, find the unknown node voltages and the current through R4 using Multimeters.

Problem 3-10: For the circuit of Problem 3-1, use the indicators to show that Kirchhoff's voltage law holds for all three loops of this circuit.

Problem 3-11: For the circuit of Problem 3-1, use the indicators to show that the current through R1 equals the current through R2 and R3 (verifying Kirchhoff's current law).

Problem 3-12: For the circuit shown below, find the unknown node voltages and the current through R3 using indicators.

Problem 3-13: For the circuit of Problem 3-12, find the unknown node voltages and the current through R3 using Multimeters.

Problem 3-14: For the circuit of Problem 3-12, find all node voltages using the SPICE DC Operating Point Analysis.

Problem 3-15: For the circuit of Problem 3-12, find the power absorbed by R5 using the power meter.

Problem 3-16: For the circuit of Problem 3-12, find the current through all resistors using the indicators.

Problem 3-17: For the circuit of Problem 3-12, find the power absorbed by R4 using the power meter.

Problem 3-18: For the circuit of Problem 3-12, find the power supplied by I1 using the power meter.

Problem 3-19: For the circuit of Problem 3-12, find the unknown node voltages and the current through R2 using indicators.

Problem 3-20: For the circuit of Problem 3-12, find the unknown node voltages and the current through R4 using Multimeters.

Problem 3-21: For the circuit of Problem 3-12, use the indicators to show that Kirchhoff's voltage law holds for all three loops of this circuit.

Problem 3-22: For the circuit of Problem 3-12, use the indicators to show that the current through I1 equals the current through R1 and R2 (verifying Kirchhoff's current law).

Problem 3-23: For the circuit shown below, find the unknown node voltages and the current through R3 using indicators.

Problem 3-24: For the circuit of Problem 3-23, find the unknown node voltages and the current through R3 using Multimeters.

Problem 3-25: For the circuit of Problem 3-23, find all node voltages using the SPICE DC Operating Point Analysis.

Problem 3-26: For the circuit of Problem 3-23, find the power absorbed by R5 using the power meter.

Problem 3-27: For the circuit of Problem 3-23, find the current through all resistors using the indicators.

Problem 3-28: For the circuit of Problem 3-23, find the power absorbed by R4 using the power meter.

Problem 3-29: For the circuit of Problem 3-23, find the power supplied by V1 using the power meter.

Problem 3-30: For the circuit of Problem 3-23, find the unknown node voltages and the current through R2 using indicators.

Problem 3-31: For the circuit of Problem 3-23, find the unknown node voltages and the current through R4 using Multimeters.

Problem 3-32: For the circuit used in section 3.C, find the diode voltage and current when the diode saturation current is 10^{-13}amp, 10^{-14}amp, 10^{-15}amp, 10^{-16}amp. By hand, plot the diode current versus the saturation current, and plot the diode voltage versus the saturation current.

Problem 3-33: Find the diode current and voltage in the circuit below.

Problem 3-34: Find the diode current and voltage in the circuit below.

Problem 3-35: Find the diode current and voltage in the circuit below.

Problem 3-36: Find the diode current and voltage in the circuit below.

Problem 3-37: Find the diode current and voltage in the circuit below.

Problem 3-38: Find the diode current and voltage in the circuit below.

Problem 3-39: Find the Thevenin equivalent circuit for the circuit below at the indicated terminals:

Problem 3-40: Find the Thevenin equivalent circuit for the circuit below at the indicated terminals:

Problem 3-41: Find the Thevenin equivalent circuit for the circuit below at the indicated terminals:

Problem 3-42: Find the Thevenin equivalent circuit for the circuit below at the indicated terminals:

Problem 3-43: Find the Thevenin equivalent circuit for the circuit below at the indicated terminals:

Problem 3-44: Find the Thevenin equivalent circuit for the circuit below at the indicated terminals:

Problem 3-45: Find the Thevenin equivalent circuit for the circuit below at the indicated terminals:

Problem 3-46: Find the transistor operating point I_C and V_{CE} in the circuit below:

Problem 3-47: Find the transistor operating point I_D and V_{DS} in the circuit below:

Problem 3-48: Find the transistor operating point I_C and V_{CE} in the circuit below:

Problem 3-49: Find the transistor operating point I_D and V_{DS} in the circuit below:

Hint: You may want to add a zero volt DC source in series with the drain as shown below so that you can easily measure the drain current:

Problem 3-50: Find the transistor operating point I_C and V_{CE} in the circuit below:

Problem 3-51: Find the transistor operating point I_D and V_{DS} in the circuit below:

PART 4
DC Sweep

The DC Sweep can be used to find all DC voltages and currents in a circuit. The DC Sweep is similar to the analyses of Part 3 but adds more flexibility. The added flexibility is the ability to allow DC sources to change voltages or currents. For example, the circuit on page 115 will give us results only for the single value of Vx = 15 V if the DC Operating Point Analysis is used. For each different value of Vx we are interested in, we must run the simulation again. If we use the DC Sweep, we can simulate the circuit for several different values of Vx in the same simulation. How node voltages vary for changing source voltages or how a BJT's bias collector current changes for different DC supply voltages would be example applications of the DC Sweep. As in the node voltage analysis, all capacitors are assumed to be open circuits, and all inductors are assumed to be short circuits.

An important note about this section is that the simulations we perform can only be done using the DC Sweep analysis. In the previous section, we found that we could use the DC Operating Point simulation to get the same results as we found using the current and voltage indicators, or the Multimeter. In the AC Analysis section, we will see that we can use the AC Sweep analysis or the Bode plotter to view similar results. In the Time Domain section, we can use the Oscilloscope or the Transient Analysis to view similar results. In most of the other sections, we can either run an analysis and plot results with the Postprocessor, or view results using one of the instruments. The DC Sweep simulation differs in that there is no standard instrument that will allow us to view the same results as we can with the DC Sweep analysis. Thus, in this section we will not demonstrate similar simulations using both the analyses and the instruments to view results. Instead, we will only run the DC Sweep analysis.

4.A. Basic DC Analysis

We will start with a modification of the circuit discussed in section 3.A:

The question we will ask is: How does the voltage at **Vo** vary as **Vx** is raised from 0 to 25 volts? We will also view some of the currents through the components. Since this is a DC Sweep, all capacitors are assumed to be open circuits, and all inductors are assumed to be short circuits. We will now set up the DC Sweep. From the menu bar select **Simulate**, **Analyses**, and then **DC Sweep**:

168 DC Sweep EWB Multisim

There is only one DC source in our circuit, so Multisim selects this source as the one to be swept. Note that we called the source **Vx** while Multisim refers to the source as **vvx**. This is done because the SPICE convention for naming voltage sources is that the name must start with the letter v. In Multisim, you can have a voltage source start with any letter because Multisim attaches the letter v as a prefix for voltage sources. The adding of prefixes applies to other components such as resistors, capacitors, inductors, and current sources. We would like to sweep **Vx** from 0 volts to 25 volts in 0.01 volt increments. Fill in the dialog box as shown:

Note that we are setting the parameters for sweeping **Source 1**. Since we only have a single source this is all we can sweep. However, we are able to set up sweeps for two sources so that we can run nested sweeps. See section 4.D for an example of a nested sweep.

The parameters for the sweep are now set up. Before we can simulate the circuit we must specify output variables to plot during the simulation. Click on the **Output variables** tab:

We must select at least one variable to plot during the simulation. See Part 2 for information on displaying traces with the Postprocessor. When the simulation is complete, we can display traces that incorporate any node voltage in the circuit.

Displaying currents, however, is a bit more difficult. Here, we will illustrate how to make the current through a few of the resistors available for plotting using the Postprocessor.

First, we will select the voltage at node **Vo** to be plotted during the simulation. We need to select at least one variable to be plotted during the simulation. If we are interested in other variables after the simulation is complete, we can display them using the Postprocessor. Click the *LEFT* mouse button on the text **Vo** as shown:

Click on the **Plot during simulation** button.

 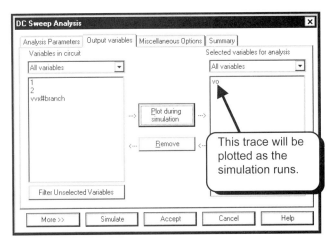

Vo will now be plotted as the simulation runs.

We are now ready to run the simulation. Click on the **Simulate** button. The **Analysis Graphs** window will open and display the selected output voltage:

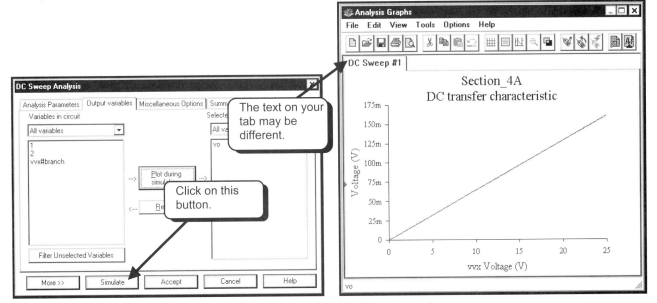

The graph plots the selected variable versus the sweep variable, **Vx** in this example.

We can now use the Postprocessor to view other voltages in the circuit. To access the Postprocessor, select **Simulate**, and then **Postprocess** from the Multisim menus:

Note that in the **Analysis Results** section, all analyses you have run in this Multisim session are listed. In my case, I have only run a single analysis, this DC sweep. If you have run the simulation more than once, or have run other simulations before you ran the simulation for this section, your dialog box will list more than the single analysis shown in my dialog box. The analysis we just ran should be selected. If not, select it before continuing.

We can now plot the traces in which we are interested. First, we would like to place these traces on a new plot, so click on the **New Page** button:

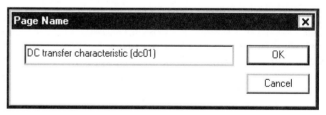

The name we place here will be the name that appears on the tab of the graph when displayed in the Analysis Graphs window. Give the page a name and click on the **OK** button:

We notice that we can only plot node voltages such as **v(1)**, **v(2)**, **vo**, and the voltage of the DC source **Vx** (indicated by **vvx** in this dialog box). The only current listed is current through voltage sources. The line **vvx#branch** specifies the branch current through voltage source **Vx**. We note that the currents through the resistors are not listed. To view these, we need either to add a current monitor to the circuit (a current-dependent voltage source) or to specify the current during the analysis setup. We will show both later in this section. For now we are limited to the node voltages and the current through the DC source.

Double-click the *LEFT* mouse button on the text **v(1)** to add the trace to the list:

Click on the **New Graph** button:

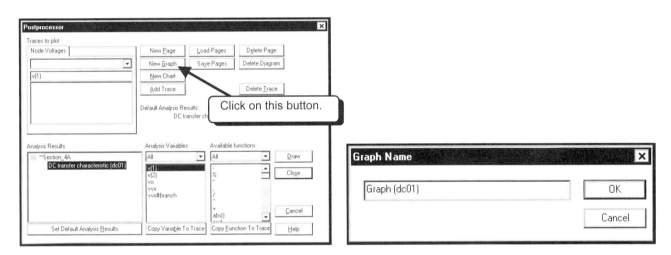

The name we enter here will be the title of the plot. Enter the text `Node Voltages` and click on the **OK** button:

Next, we need to make the selected node voltage a trace on the graph. Click on the **Add Trace** button:

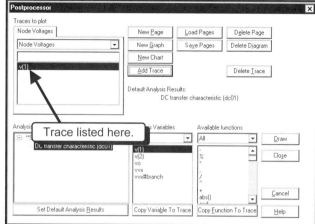

The trace **v(1)** will now be added to the graph titled **Node Voltages**.

We will now repeat the procedure to add the trace v(2) to the same graph on the same page. Double-click the *LEFT* mouse button on the text **v(2)** to add the trace to the list

Click on the **Add Trace** button:

The trace **v(2)** will now be added to the graph titled **Node Voltages**. To view the graph, click on the **Draw** button:

To see the Analysis Graphs window, close the **Simulation Error Log/Audit Trail** window and close the **Postprocessor** window and bring the **Analysis Graphs** window to the front:

On my plot, a portion of the trace appears to be drawn incorrectly. When I create graphs with a single trace, the line shown above does not appear:

This error will be fixed in a future release of Multisim.

To show currents, we must either add a current monitor or specify the current traces as outputs when we set up the DC Sweep analysis. We will show both here. Close the **Analysis Graphs** window. Place a part called a **Current Controlled Voltage Source**. This part is located in the **Sources** toolbar:

We would like to monitor the current through **R1**. To add the monitor, we will have to break the circuit and insert the current monitor between **R1** and node **2** as shown:

The current-dependent voltage source is named **V1**. Its gain is **1 Ohm**. This means that the output voltage of the source is 1 ohm times the current it measures, or the transfer ratio of this current monitor is 1 volt per amp. Also note that we added a resistor in parallel with the source. The value and name of this are not important. It was added so that we can easily name the node between the two elements. In my circuit the node is named **IR1**. We need to name the node so that we can easily view its voltage. To name the node, double-click the *LEFT* mouse button on the wire between the resistor and the voltage source.

EWB Multisim	DC Sweep	175

When you have finished changing the circuit, select **Simulate**, **Analyses**, and then **DC Sweep**:

We will now show a second way of viewing the current through R1. Click on the **Output variables** tab:

Click on the **More** button:

Click on the **Add device/model parameter** button.

We would like to view the current through R1. We need to specify the **Device Type** as a resistor and then select R1. Click the *LEFT* mouse button on the down triangle ▼ for the **Device Type** field:

Select **Resistor**:

In the **Name** field of my dialog box, resistor R1 is already selected (**rr1**). If your dialog box does not specify resistor R1, click the *LEFT* mouse button on the down triangle ▼ for the **Name** field and select **rr1**. The last thing we must do is specify what parameter of the resistor we would like to plot. Click the *LEFT* mouse button on the down triangle ▼ for the **Parameter** field to display the available parameters:

Select current by clicking on the lower case **i**. Be careful not to select **L**, the length of the resistor. If you select the correct parameter, the text **Current** will be displayed in the **Description** field:

EWB Multisim   DC Sweep   177

Click on the **OK** button. The resistor current will now be displayed as a variable we can plot:

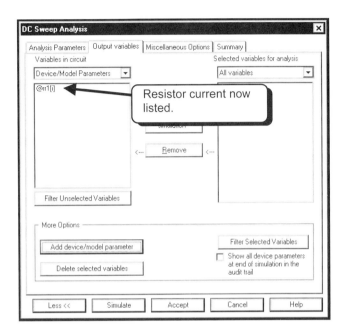

We must now add this trace to the traces that are plotted during the simulation. Click the *LEFT* mouse button on the text **@rr1[i]** to select it:

Click on the **Plot during simulation** button:

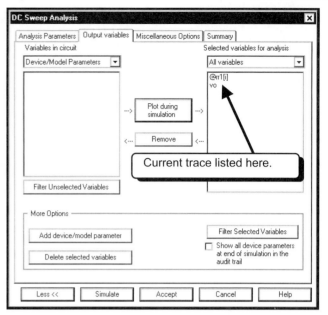

The trace will now be displayed as the simulation is running, and we can also plot it with the Postprocessor as well.

Click on the **Simulate** button to simulate the circuit. The voltage at node **Vo** and the current through **R1** will be displayed in the **Analysis Graphs** window:

We will now use the Postprocessor to plot the current traces. To access the Postprocessor, select **Simulate** and then **Postprocess** from the Multisim menus:

The data set currently selected is from the previous DC Sweep. We need to select the data set from the analysis we just ran. Click the **LEFT** mouse button on the ⊞ as shown. The tree will expand to show the available data sets:

Click on the text **DC transfer characteristic (dc02)** to select the data set:

 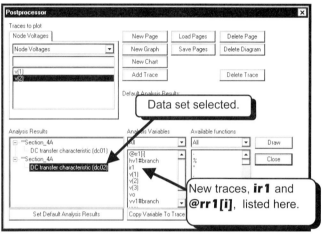

We see that the voltage at the sensor output (**ir1**) and the parameter for the resistor we added (**@rr1[i]**) are now listed. Use the procedure shown previously to add a new page, and then add a graph called `Resistor Currents`. Plot both current traces, **ir1** and **@rr1[i]**. My plot is shown below:

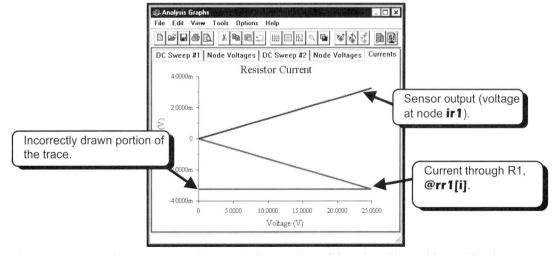

We see that the two traces are the same except that one is the negative of the other. The problem with using a parameter such as **@rr1[i]** is that we do not know which direction Multisim considers as the positive direction of current. It turns out that if I rotated R1 in the schematic by 180 degrees, the current would then match up with the sensor output. We also know that the y-axis for the current **@rr1[i]** should be current rather than voltage. We can easily change the label of the axis if we wish. This procedure was covered in section 2.F.3. With a sensor we know which direction is positive (left to right in my schematic) because the arrow in the part graphic indicates the positive direction.

Since I know that the current through R1 is left to right, I can always plot the trace -@rr1[i] to correct for the sign error:

We see that the two methods produce the same result.

EXERCISE 4-1: Find the voltages at nodes **1**, **2**, and **3** if the source voltage **V1** is swept from a DC voltage of 6 volts to a DC voltage of 36 volts.

SOLUTION: The results of the DC Sweep are shown graphically using Probe:

4.B. Diode I-V Characteristic

We would now like to use SPICE to obtain the I-V characteristic of a semiconductor diode. Wire the circuit shown below:

EWB Multisim   DC Sweep   181

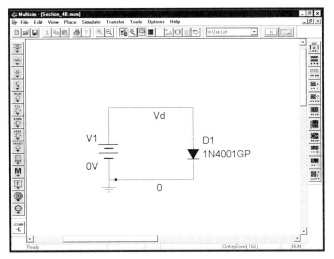

Note that we labeled the top node as **Vd** rather than a number. To label a wire, double-click on the wire.[1] The diode is located in the **Diodes** toolbar:

Select the **1N4001GP** diode model:

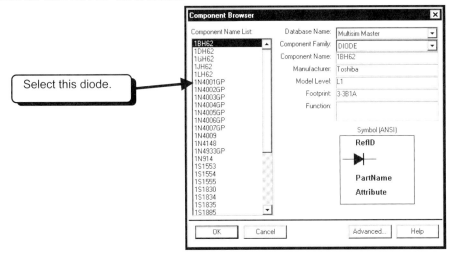

After creating the circuit, we need to set up a DC sweep. Select **Simulate**, **Analyses**, and then **DC Sweep** and fill in the dialog box as shown:

---

[1] See section 1.G to label a wire or give a node a name.

The settings in the **DC Sweep Analysis** dialog box specify that voltage source **V1** is to be swept from **-15** volts to **1** volt in **0.001** volt steps.

To view the output current, we need to add the diode current to the list of output variables. Click the *LEFT* mouse button on the **Output variables** tab:

Click on the **More** button and then select the diode current as an output variable. See pages 175-178 for a detailed explanation of how to specify a device current as an output variable.

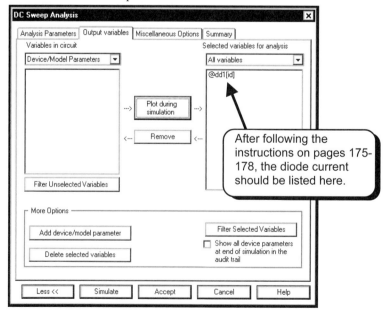

Since the diode current is listed as an output variable, it will be plotted automatically when the simulation runs.

EWB Multisim	DC Sweep	183

Click on the **Simulate** button to run the simulation. The **Analysis Graphs** window should pop up and display the I-V characteristic:

To modify the plot settings, click the ***RIGHT*** mouse button on the graph:

Select **Properties**:

Change the properties of the plot so that it looks like the plot below:

Note that the diode current shown here is too large for this device to handle. A 1N4001 can only handle 1 amp of current or less. The reason the current goes to nearly 2 amps is because we applied a large voltage across the diode, 1 V or more, and this is the current the model produces with a 1 volt diode drop. For a typical 1N4001, the diode voltage should be closer to 0.7 volt.

EXERCISE 4-2: Display the I-V characteristic of a green LED:

SOLUTION: Use the DC Sweep analysis and sweep the voltage from 0 to 2.5 volts. After plotting the trace, modify the y-axis to only display currents from 0 to 1 amp.

4.C. DC Transfer Curves

One of the more useful functions of the DC Sweep is to plot transfer curves. A transfer curve plots an output versus an input. A DC transfer curve plots an output versus an input, assuming all capacitors are open circuits and all inductors are short circuits. In a DC Sweep, all capacitors are replaced by open circuits and all inductors are replaced by short circuits. Thus the DC Sweep is used for DC transfer curves.

4.C.1. Zener Clipping Circuit

The circuit below should clip positive voltages at the Zener breakdown voltage and negative voltages at the diode cut-in voltage, approximately 0.6 to 0.8 volt:

We would like to sweep **Vs** from −15 volts to +15 volts and plot the output. Select **Simulate**, **Analyses**, and then **DC Sweep** and fill in the dialog box as shown:

Next, select the **Output variables** tab and plot **Vo** during the simulation:

Click on the **Simulate** button to run the simulation. The result will be displayed in the **Analysis Graphs** window:

I manually modified the x- and y-axis titles. We see that this circuit limits the output at the Zener breakdown voltage of 5.6 volts and –0.7 volt, minus the diode turn-on voltage.

EXERCISE 4-3: Find the transfer curve of the circuit below:

SOLUTION: Sweep **V1** from –15 V to +15 V as in the previous example and specify Vo as the variable to plot during the simulation.

Notice that the output is zero for input voltages from approximately –0.7 V to about 5.6 V.

4.C.2. NMOS Inverter Transfer Curve

We would like to plot the transfer curve **Vo** versus **Vin** for the NMOS inverter below:

This circuit is an NMOS inverter with a depletion MOSFET load. The driver transistor is an N-type enhancement MOSFET and the load transistor is an N-type depletion MOSFET. In SPICE, both use the same model parameters. The way SPICE distinguishes between enhancement and depletion devices is that the threshold voltage is positive for enhancement devices and negative for depletion devices. Both types of transistors use the SPICE NMOS model. In the screen capture above, Multisim uses different symbols to emphasize that one is a depletion type and the other is an enhancement type. To create the circuit above, you need to place a part called **MOS_3TDN_VIRTUAL** and one called **MOS_3TEN_VIRTUAL**. Both are located in the Transistors toolbar.

Before we create the models for the two transistors, we need to discuss the basic operation of the devices and how the standard SPICE equations may differ from convention. Many textbooks describe MOSFET operation by the following equations:

$$I_D = K(V_{GS} - V_T)^2 (1 + \lambda V_{DS})$$ saturation region

$$I_D = K(2(V_{GS} - V_T)V_{DS} - V_{DS}^2)(1 + \lambda V_{DS})$$ linear region

For our example, the load MOSFET is a depletion-mode NMOS transistor. Its parameters are K = 20 µA/V², V_T = –1.5 V, and λ = 0.05 V⁻¹. The driver is an enhancement-mode NMOS transistor. Its parameters are K = 20 µA/V², V_T = +1.5 V, and λ = 0.05 V⁻¹. We note that the only difference between the two is the threshold voltage, V_T.

First, we need to create models to describe these transistors. The standard SPICE MOSFET model uses the following equations to describe MOSFET operation:

$$I_D = \frac{K_P}{2}(V_{GS} - V_{TO})^2 (1 + \lambda V_{DS})$$ saturation region

$$I_D = \frac{K_P}{2}(2(V_{GS} - V_{TO})V_{DS} - V_{DS}^2)(1 + \lambda V_{DS})$$ linear region

The conversion between the two sets of equations is shown in Table 4-1. The SPICE model parameter names are close to the standard names used to represent MOSFET operation in many textbooks. One difference is that the SPICE model parameter Kp is twice the value of K. Thus, in our model we should set K_P = 40 µA/V². All other model parameters will be the same.

After drawing the circuit, the next thing we need to do is modify the models used by the virtual components. First, double-click on the driver MOSFET graphic to change its properties:

Table 4-1	
SPICE Model Parameter	Equation Variables
Kp	2K
VTO	V_T
lambda	λ

Select the **Value** tab if it is not already selected. To modify the SPICE model, click the *LEFT* mouse button on the **Edit Model** button:

We are not allowed to change the name of the model, but we can change the values of the model parameters. Change the value of **VTO** to 1.5, **KP** to 40e-6, and **LAMBDA** to 0.05:

Click on the **Change Part Model** button to apply the changes to the part we double-clicked on.

Click on the **OK** button to return to the schematic:

Note that the model name now has an asterisk next to it indicating that the model it uses has been changed. Use the same procedure to change the model for the load transistor. For that transistor, change the value of **VTO** to –1.5, **KP** to 40e-6, and **LAMBDA** to 0.05. Below I show the model editor window and the resulting schematic:

Once again, the asterisk on the schematic indicates that the model has been modified.

We are now ready to set up the DC Sweep. Select **Simulate**, **Analyses**, and then **DC Sweep** and fill in the dialog box as shown. The dialog box specifies that **Vin** should be swept from **0** volts to **5** volts in 1 mV steps. Under the **Output variables** tab, we have selected to display the voltage at node **Vo**:

Click on the **Simulate** button to run the simulation. In the screen capture below, I have modified the axis labels and added a grid to the plot:

EXERCISE 4-4: Find the transfer curve for a CMOS inverter. The top transistor is a PMOS enhancement device, and the bottom transistor is an NMOS enhancement device. For the NMOS transistor, let Kp = 24 µA/V^2, Vto = 1.5 V, and lambda = 0.01 V^{-1}. For the PMOS transistor, let Kp = 24 µA/V^2, Vto = –1.5 V, and lambda = 0.01 V^{-1}. In most processes, Kp for PMOS devices is less than Kp for NMOS devices, and the difference between the two values is equalized by changing the width-to-length ratio of the PMOS and NMOS devices to make the effective value of Kp the same. The effective value of Kp is $\left(\dfrac{W}{L}K_P\right)$. This Multisim model does not let us specify a width-to-length ratio, so we will just specify that the Kp values for both transistors are the same.

SOLUTION: Create the circuit below:

Create the following models for the two transistors:

Run a DC Sweep to sweep **Vin** from 0 to 5 V in 0.001 V increments. A small increment is required for a CMOS inverter because the slope of the transfer curve is very steep. Specify **Vo** as the output variable:

4.D. Nested DC Sweep — BJT Characteristic Curves

The nested DC Sweep can be used to generate characteristic curves for transistors. We will illustrate generating these curves using a BJT. Wire the circuit below:

We need to set up a nested DC sweep. The inner loop will sweep VCE from 0 to 15 volts, and the outer loop will step IB from 0 to 100 µA in 10 µA steps. Select **Simulate**, **Analysis**, and then **DC Sweep** from the Multisim menus and fill in the dialog box as shown:

The way this dialog box is set up, for each value of source **IB**, **VCE** will be swept from 0 volts to 15 volts in 10 mV steps. In other words, **IB** will be set to 0 A and then **VCE** will be swept from 0 volts to 15 volts; then **IB** will be set to 10 µA and **VCE** will be swept from 0 volts to 15 volts; then **IB** will be set to 20 µA and **VCE** will be swept from 0 volts to 15 volts, and so on. This nested DC sweep will contain eleven I_C versus V_{CE} curves.

The next thing we must do is specify the output variable. We are interested in the collector current I_C of **Q1**. We can view this current in two ways. (1) We can use the procedure on pages 175-178 to add the collector current of **Q1**, which is referred to as a device parameter. (2) The second way is to note that the collector current is also equal to the current supplied by the source **VCE**. The definition of positive current for a voltage source is that current is positive if it flows into the positive voltage terminal. This direction is the opposite of the definition of positive collector current. Thus, to plot the collector current, we can specify the output variable to be the negative of the current through source **VCE**. We cannot specify functions until we use the Postprocessor, so we cannot plot the negative of the source current during the simulation. We would have to run the simulation and then generate a new plot with the Postprocessor. I will plot the collector current of Q1 since this is the correct direction for the current. Select the **Output variables** tab and follow the procedure on pages 175-178 to add the collector current of **Q1**:

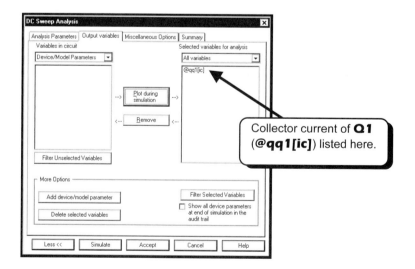

Collector current of **Q1** (**@qq1[ic]**) listed here.

Click on the **Simulate** button to run the simulation and plot the collector current. In the screen capture below, I have modified the titles and added a grid after the simulation was complete:

This is the characteristic family of curves for a 2N3904 NPN bipolar junction transistor.

EXERCISE 4-5: Use the nested DC Sweep to find the characteristic curves of a 2N3821 jFET.

SOLUTION: Draw the circuit below:

For each value of **VGS**, we want to sweep **VDS** from 0 to 15 volts. Thus, we want **VDS** to be the inner loop or source 1 and **VGS** to be the outer loop or source 2. Note also that for jFETs we must sweep **VGS** from zero to a negative value. Fill in the DC Sweep dialog box as shown:

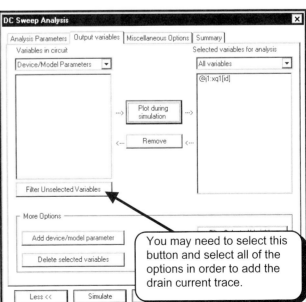

You may need to select this button and select all of the options in order to add the drain current trace.

Note that the model for this jFET is a subcircuit. In order for the drain current to appear in the left pane of the right screen capture above, you will need to click on the **Filter Unselected Variables** button and select all of the options.

Simulate the circuit and modify the labels to obtain the plot below:

4.E. DC Current Gain of a BJT

In this section we will investigate how the DC current gain (H_{FE}) of a bipolar junction transistor varies with DC bias collector current I_{CQ} and DC bias collector-emitter voltage V_{CEQ}. We will use the basic circuit shown below for all simulations:

In this circuit we are making the approximations $I_C \cong I_E$ and $V_{CE} = V_1 + V_{BE} \approx V_1$. We are making these approximations because we cannot directly sweep the collector current. The displayed curves will have the same shape as if we were sweeping the collector current.

4.E.1. H_{FE} Versus Emitter Current

Our first analysis will display how H_{FE} of a transistor varies with DC emitter current.[2] In many applications we want to use a transistor where it has its maximum current gain. This plot will tell us how to choose the emitter current for maximum current gain. We can then bias the transistor at this current. This plot is easily generated with a DC Sweep. We will generate this curve with V_{CE} constant at approximately 5.7 V (assuming the $V_{BE} \approx 0.7$ volts). Note in the circuit above that V_{CE} is held approximately constant by the DC Source V1, and we will not change it during the simulation. V_{BE} will change slightly as the current changes, but we will assume that this change is small. Wire the circuit above and then select **Simulate**, **Analyses**, and then **DC Sweep**. Fill in the dialog box as shown:

---

[2]Thanks to John D. Welkes of Arizona State University for this example.

EWB Multisim   DC Sweep   195

We are sweeping the emitter current from 100 μA to 100 mA in 10 μA steps.

To plot the DC current gain, H_{FE}, we need to plot the ratio of I_C/I_B using the Postprocessor. Since current data is not normally available to the Postprocessor, we need to specify the traces in the **Output variables** portion of the setup. Select the **Output variables** tab and then follow the procedure on pages 175-178 to add the base current and collector current as traces that will be plotted during the simulation:

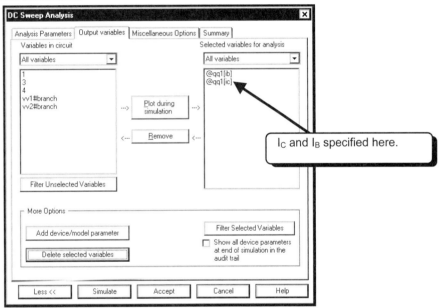

Click on the **Simulate** button to run the simulation. The **Analysis Graphs** window will open and display the base current and collector current:

This graph shows that the simulation ran, but it is not the plot we are after. To generate a plot of I_C/I_B, we need to use the Postprocessor. Select **Simulate** and then **Postprocess** from the Multisim menus:

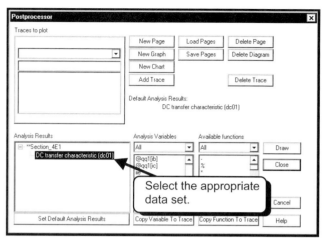

If you have run several simulations before this one, you may have several data sets listed in the **Analysis Results** pane. You will need to select the appropriate data set before continuing. We wish to create a new page containing a graph of I_C/I_B. See sections 2.A to 2.C on creating new plots with the Postprocessor. Add a plot of I_C/I_B. If you have only run this simulation since you started Multisim, your dialog box will look as shown:

Your trace, **@qq1[ic]/@qq1[ib]**, will be slightly different if you had more than one data set available. Click on the **Draw** button to plot the trace. In the screen capture below, I have modified the labels slightly:

This is not the plot we are really interested in. We would actually like the x-axis to be a log scale. To change it, click the **RIGHT** mouse button on the graph:

EWB Multisim   DC Sweep   197

Select **Properties**:

Select the **Bottom Axis** tab if it is not already selected:

In the screen capture above, the x-axis (bottom axis) is specified as **Linear**. We need to change this axis to a logarithmic scale, so click the *LEFT* mouse button on the circle ○ next to the text **Logarithmic**. The circle should fill with a dot ⊙, indicating that the option is selected:

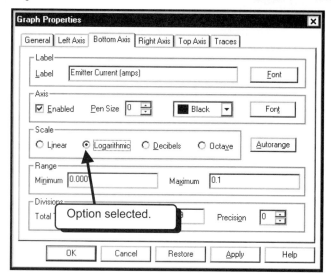

Click on the **OK** button to see the results:

If we turn on the cursors, we see that the maximum current gain is **170.4557** and occurs at an emitter current of about 10 mA:

In our analysis, we made the assumption that V_{BE} was approximately constant. We will now plot the base-emitter voltage to see if this was a good approximation. V_{BE} is equal to zero minus the voltage at node **4** in our circuit:

We see that V_{BE} varies by about 200 mV as the emitter current changes from 100 µA to 100 mA. We see that V_{BE} is roughly constant for the simulation, and therefore V_{CE} is roughly constant as well since V_{CE} is equal to **V1** minus V_{BE}.

EXERCISE 4-6: For the 2N3906 PNP bipolar junction transistor, find the collector current where H_{FE} is at its maximum value. Display the maximum value of H_{FE} and the emitter current where it occurs. Let V_{EC} be constant at about 5.7 V.

SOLUTION: Use a circuit similar to the one used in the previous example, but use the 2N3906 transistor:

This curve appears to be identical to the one for the 2N3904. This is expected since the 2N3906 is the PNP complement of the 2N3904.

4.E.2. H_{FE} Versus I_C for Different Values of V_{CE}

The next thing we would like to do is to see how the H_{FE} versus I_E curve is affected for different values of DC collector-emitter voltage. The curve in the previous example was generated at $V_{CE} \cong 5.7$ V. We would now like to generate four curves at different values of V_{CE} and plot them all on the same graph. We will generate curves at $V_{CE} \cong 5$ V, 10 V, and 15 V. We will use the same circuit as in the previous section:

Select **Simulate**, **Analyses**, and then **DC Sweep** to bring up our previous sweep setup:

The dialog box should retain its settings from the previous simulation, so all we need to do is specify the nested sweep information. We need to specify **source 2** as the source that controls VCE (in my case this is **V1**). Fill in the second source's information as shown:

The rest of the setup, including the **Output variables** information, should be the same as in the previous run and should be retained from the last simulation. If not, set up the **Output variables** tab the same as you did in section 4.E.1. Run the simulation and then open the Postprocessor:

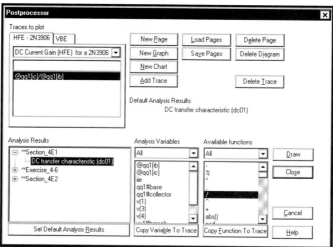

My dialog box shows three sets of data because I have run three simulations since I started using Multisim in this session. For my most recent simulation I have been using the circuit named **Section_4E2**, so I will need to select this data. I must first expand the tree to select the data. Click on the ⊞ next to the text **Section_4E2**:

DC Sweep

 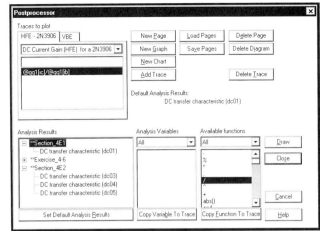

Click on one of the data sets to select it:

 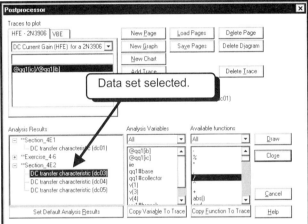

Create a new page by clicking on the **New Page** button:

 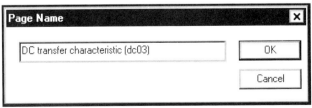

Name the page `HFE vs IE and VCE` and then click on the **OK** button:

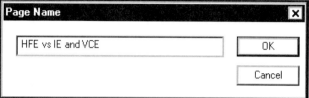

Click on the **New Graph** button and enter the name `Current Gain (HFE) Versus IE and VCE`:

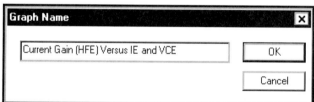

Click on the **OK** button:

Next, create the trace `dc03.@qq1[ic]/dc03.@qq1[ib]`:

Note that in my dialog box above, I have run several simulations in the current Multisim session. The data I am plotting is from the third DC Sweep, and the trace expression I am entering contains the text **dc03** to specify that I am plotting the data from the third simulation. Your expression may be slightly different in that it plots data from a different numbered sweep. Click on the **Add Trace** button to add the trace to the list:

The trace we just added was for a single value of V_{CE}. We would like to generate a plot with all three values of VCE, so we need to add three more traces. Click the **LEFT** mouse button on the text as shown below to select the second value of V_{CE} for our simulation:

Next, create the trace `dc04.@qq1[ic]/dc04.@qq1[ib]`:

Note that this trace is the same as the previous one except that it contains the text **dc04**, which specifies the fourth DC sweep that I have run. This is actually the second DC sweep for this simulation, which has three, one at $V_{CE} \cong 5$ volts, one at $V_{CE} \cong 10$ volts, and one at $V_{CE} \cong 15$ volts. Click on the **Add Trace** button to add the trace to the list:

Repeat the procedure to add the third trace for $V_{CE} \cong 15$ volts:

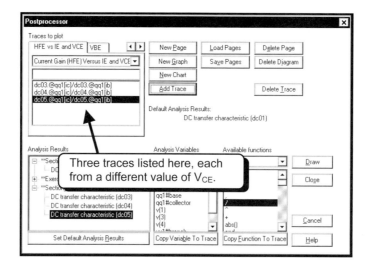

Click on the **Draw** button to create the plot:

Change the labels to correctly annotate the plot, and change the x-axis to a log scale:

We see that the value of V_{CE} does affect the value of the transistor's current gain, H_{FE}. What we realize is that our bias point specifies the collector current (I_C) and collector-emitter voltage (V_{CE}), and these two parameters affect the performance of the device. If we wanted to bias this transistor for maximum H_{FE}, the curve above could be used as a guide for how to choose our bias values of V_{CE} and I_C.

4.F. Problems

Problem 4-1: For the circuit below, plot the voltage Vo versus Vin as Vin is varied from –10 volts to 15 volts.

Problem 4-2: For the circuit of Problem 4-1, plot the voltage V1 versus Vin as Vin is varied from 0 volts to 15 volts.

Problem 4-3: For the circuit of Problem 4-1, plot the voltage V2 versus Vin as Vin is varied from 0 volts to 25 volts.

Problem 4-4: For the circuit of Problem 4-1, plot the sum of voltages V1 and V2 versus Vin as Vin is varied from 0 volts to 25 volts.

Problem 4-5: For the circuit of Problem 4-1, plot the voltage across resistor R3 versus Vin as Vin is varied from 0 volts to 25 volts.

Problem 4-6: For the circuit of Problem 4-1, plot the ratio Vo/Vin versus Vin as Vin is varied from 0 volts to 7 volts.

Problem 4-7: For the circuit of Problem 4-1, plot the current through resistor R1 versus Vin as Vin is varied from −8 volts to 16 volts.

Problem 4-8: For the circuit of Problem 4-1, plot the current through resistor R3 versus Vin as Vin is varied from −10 volts to 10 volts.

Problem 4-9: For the circuit of Problem 4-1, plot the power absorbed by resistor R1 versus Vin as Vin is varied from −25 volts to 25 volts.

Problem 4-10: For the circuit of Problem 4-1, plot the power supplied by voltage source Vin versus Vin as Vin is varied from −25 volts to 25 volts.

Problem 4-11: For the circuit below, plot the voltage Vo versus I1 as I1 is varied from −10 mA to 15 mA.

Problem 4-12: For the circuit of Problem 4-11, plot the voltage V1 versus I1 as I1 is varied from 0 mA to 15 mA.

Problem 4-13: For the circuit of Problem 4-11, plot the voltage V2 versus I1 as I1 is varied from 0 mA to 25 mA.

Problem 4-14: For the circuit of Problem 4-11, plot the sum of voltages V1 and V2 versus I1 as I1 is varied from 0 mA to 25 mA.

Problem 4-15: For the circuit of Problem 4-11, plot the voltage across resistor R3 versus I1 as I1 is varied from 0 mA to 25 mA.

Problem 4-16: For the circuit of Problem 4-11, plot the ratio Vo/Vin versus I1 as I1 is varied from 0 mA to 7 mA.

Problem 4-17: For the circuit of Problem 4-11, plot the current through resistor R1 versus I1 as I1 is varied from −8 mA to 16 mA.

Problem 4-18: For the circuit of Problem 4-11, plot the current through resistor R3 versus I1 as I1 is varied from −10 mA to 10 mA.

Problem 4-19: For the circuit of Problem 4-11, plot the power absorbed by resistor R1 versus I1 as I1 is varied from −25 mA to 25 mA.

Problem 4-20: For the circuit of Problem 4-11, plot the power supplied by current source I1 versus I1 as I1 is varied from −25 mA to 25 mA.

Problem 4-21: For the circuit below, plot the voltage Vo versus Vin as Vin is varied from −10 volts to 15 volts.

Problem 4-22: For the circuit of Problem 4-21, plot the voltage V1 versus Vin as Vin is varied from 0 volts to 15 volts.

Problem 4-23: For the circuit of Problem 4-21, plot the voltage V2 versus Vin as Vin is varied from 0 volts to 25 volts.

Problem 4-24: For the circuit of Problem 4-21, plot the sum of voltages V1 and V2 versus Vin as Vin is varied from 0 volts to 25 volts.

Problem 4-25: For the circuit of Problem 4-21, plot the voltage across resistor R4 versus Vin as Vin is varied from 0 volts to 25 volts.

Problem 4-26: For the circuit of Problem 4-21, plot the ratio Vo/Vin versus Vin as Vin is varied from 0 volts to 7 volts.

Problem 4-27: For the circuit of Problem 4-21, plot the current through resistor R2 versus Vin as Vin is varied from −8 volts to 16 volts.

Problem 4-28: For the circuit of Problem 4-21, plot the current through resistor R3 versus Vin as Vin is varied from −10 volts to 10 volts.

Problem 4-29: For the circuit of Problem 4-21, plot the current through resistor R4 versus VA as VA is varied from −10 volts to 12 volts.

Problem 4-30: For the circuit of Problem 4-21, plot the current through resistor R5 versus VA as VA is varied from −12 volts to 10 volts.

Problem 4-31: For the circuit of Problem 4-21, plot the power absorbed by resistor R1 versus Vin as Vin is varied from −25 volts to 25 volts.

Problem 4-32: For the circuit of Problem 4-21, plot the power supplied by voltage source Vin versus Vin as Vin is varied from −25 volts to 25 volts.

Problem 4-33: Plot the I-V characteristic of a 1N914 diode. From the plot, obtain a numerical value for the diode turn-on voltage.

Problem 4-34: Plot the I-V characteristic of a 1S1830 diode. From the plot, obtain a numerical value for the diode turn-on voltage.

Problem 4-35: Plot the I-V characteristic of a 1N4740A Zener diode. From the plot, obtain a numerical value for the Zener breakdown voltage and the turn-on voltage.

Problem 4-36: Plot the I-V characteristic of a 1N4756A Zener diode. From the plot, obtain a numerical value for the Zener breakdown voltage and the turn-on voltage.

Problem 4-37: Plot the I-V characteristic of a blue LED. From the plot, obtain a numerical value for the diode turn-on voltage.

Problem 4-38: Plot the I-V characteristic of a yellow LED. From the plot, obtain a numerical value for the diode turn-on voltage.

Problem 4-39: Plot the I-V characteristic of a 1 kΩ resistor.

Problem 4-40: Plot the I-V characteristic of a 500 Ω resistor.

Problem 4-41: Plot the I-V characteristic of a 15 Ω resistor.

Problem 4-42: Plot the I-V characteristic of a 15 MΩ resistor.

Problem 4-43: Plot the transfer curve Vo versus Vin for the circuit below. Let Vin vary from −20 volts to 20 volts.

Problem 4-44: Plot the transfer curve Vo versus Vin for the circuit below. Let Vin vary from −20 volts to 20 volts.

Problem 4-45: Plot the transfer curve Vo versus Vin for the circuit below. Let Vin vary from −20 volts to 20 volts.

Problem 4-46: Plot the transfer curve Vo versus Vin for the circuit below. Let Vin vary from –20 volts to 20 volts.

Problem 4-47: Plot the transfer curve Vo versus Vin for the circuit below. Let Vin vary from –20 volts to 20 volts.

Problem 4-48: Plot the transfer curve Vo versus Vin for the circuit below. Let Vin vary from –20 volts to 20 volts.

Problem 4-49: Plot the transfer curve Vo versus Vin for the circuit below. Let Vin vary from 0 volts to 5 volts. For the driver transistor, use $Kp = 60 \ \mu A/V^2$ and $VTO = 1$ volt. For the load transistor, use $Kp = 30 \ \mu A/V^2$ and $VTO = 0.5$ volt.

Problem 4-50: Plot the transfer curve Vo versus Vin for the circuit below. Let Vin vary from 0 volts to 5 volts. For the driver transistor, use $Kp = 120 \ \mu A/V^2$ and $VTO = 1$ volt. For the load transistor, use $Kp = 30 \ \mu A/V^2$ and $VTO = 0.5$ volt.

Problem 4-51: Redo the simulation of section 4.C.2 using the circuit below. Use the same parameters for the MOSFET as used in the example. Plot Vo versus Vin.

Problem 4-52: Plot the characteristic curves for a 2N2222A NPN BJT transistor.

Problem 4-53: Plot the characteristic curves for a 2N2222A NPN BJT transistor.

Problem 4-54: Plot the characteristic curves for a 2N4401 NPN BJT transistor.

Problem 4-55: Plot the characteristic curves for a 2N3906 PNP BJT transistor.

Problem 4-56: Plot the characteristic curves for a PN2907A PNP BJT transistor.

Problem 4-57: Plot the characteristic curves for a MOSFET with the following model parameters: $Kp = 120$ $\mu A/V^2$, $VTO = 1$ V, and Lambda $= 0.001$ V^{-1}.

Problem 4-58: Plot the characteristic curves for a MOSFET with the following model parameters: $Kp = 120$ $\mu A/V^2$, $VTO = 1$ V, and Lambda $= 0.000$ V^{-1}.

Problem 4-59: Plot the characteristic curves for a MOSFET with the following model parameters: $Kp = 120$ $\mu A/V^2$, $VTO = 1$ V, and Lambda $= 0.01$ V^{-1}.

Problem 4-60: Plot the characteristic curves for a 6C5 Triode vacuum tube.

PART 5
Magnitude and Phase Simulations

Multisim provides several tools for sinusoidal steady state magnitude and phase analysis, including AC voltmeters, power meters with power factor calculation, Bode plotters, and the SPICE AC Analysis. Circuits can be analyzed at a single frequency or at multiple frequencies.

It is important to realize the difference between magnitude and phase simulations and the time domain simulations discussed in Part 6. The simulations in this section are used to find the magnitude and phase of voltages and currents. The time domain analyses are used to look at waveforms versus time. Example waveforms versus time using the Transient Analysis and the Oscilloscope instrument are:

This graph shows us a voltage versus time, and it could be represented mathematically as $5\sin(2\pi 1000t + 0°)$. The magnitude of the above waveform is 5 V and the phase is zero degrees—in phasor notation, $5\angle 0°$. All of the magnitude and phase simulations will give a result such as $5\angle 0°$ while the time domain analyses allow us to view waveforms versus time such as the above screen captures.

5.A. Magnitude and Phase Measurements at a Single Frequency

Multisim provides two ways to measure AC waveforms. The magnitude of waveforms at a single frequency can be measured using the Multimeter or the current and voltage indicators. With these tools we can only measure the magnitude of currents and voltages; we cannot measure the phase. The AC Analysis is a more general analysis tool and it can be used to find both the magnitude and phase of voltages and currents at multiple frequencies, and the results can be viewed graphically or as text output.

5.A.1. Magnitude Measurements with Instruments

The Multisim voltage and current indicators and the Multimeter allow us to easily measure RMS voltages in a circuit at a single frequency. The use of these meters is the same as in section 3.A except we must change the meter to read AC voltages rather than DC. If you leave a meter on DC, the meter will read zero or close to zero for AC voltages. As we saw with the DC measurements, the current meter is modeled as a small series resistance, and the voltage meter is modeled as a large parallel resistance. We must keep these models in mind when making measurements because they could affect our results. Unlike real AC voltmeters, these meters do not have an upper frequency limitation. From simple tests, it appears that the meters give an accurate reading anywhere from 20 Hz up to 10^{12} Hz. Thus, except for the series and parallel resistances, the meters appear to be ideal. The one limitation of the meters and indicators is that we can only measure the magnitude of voltages and currents. For AC circuits, this is only half of the answer. In order to find the phase of a voltage or a current, we need to run the SPICE AC Analysis.

We will start by measuring the magnitude of the voltages and currents in the circuit below:

Notice that the amplitude of the voltage source is **1.41** volts. This is the peak amplitude of the waveforms. We shall see that the indicators display the RMS value of voltages and currents. Remember that the RMS value is equal to the peak value divided by the square root of two: $V_{RMS} = {V_{peak}}/{\sqrt{2}}$. Since the peak value of **V1** is **1.41**, its RMS value is about 1 volt.

When we specify the value of a source, we can specify its peak value or its RMS value. To see the settings of this source, double-click the *LEFT* mouse button on the graphic of the voltage source:

For this source, I specified the RMS value to be **1.00** volt, and Multisim calculated the peak to be **1.41** volts. Also, note that the frequency is **200 Hz** and the phase is set to zero. Make the changes to your dialog box so that the settings match the ones shown and then click on the **OK** button to close the dialog box.

The next thing we need to do is to add some monitors to the circuit. As examples, we will measure the voltages at two nodes and the current through **L1**. Before doing this section, you may want to review section 3.A where we cover the use of the meters in DC applications. The use of the meters is about the same.

To place an indicator, click the *LEFT* mouse button on the Indicators button as shown below. The **Indicators** toolbar should appear:

The two buttons on the top row are the voltage and current indicators. We will first place a voltage indicator, so click the **LEFT** mouse button on the button:

The meters all perform the same function. The only difference is the arrangement of the connection terminals. Select the **VOLTMETER_V** part and click on the **OK** button. Place the part in your circuit and wire the meter to your circuit as shown:

Before we add more meters, we will see how this meter works. Click on the Run button to start the simulation:

Magnitude and Phase Simulations

The indicator jumps around a little bit, but most of the time it reads zero. This is because the default setting for the indicator is DC. The DC component of our source is zero, so the indicator is reading the correct value. Stop the simulation by clicking on the Run button again.

To change the properties of the indicator, double-click the **LEFT** mouse button on the indicator:

We see that the indicator is set to the **DC Mode**. Since we are measuring AC waveforms, we need to select the **AC Mode**. Change the **Mode** to **AC** and click on the **OK** button:

Click on the Run button to restart the simulation. After a few moments the meter should display the RMS value of the voltage it is measuring:

My meter reads 0.997 volt, which is close to the 1 V RMS setting we specified for the source. If you perform simple test measurements with the current indicator and the Multimeter, you will notice that all of the meters read the RMS value of voltages and currents rather than the peak values. Stop the simulation by clicking on the Run button again.

Next, we will measure the current through **L1** and the voltage across **R2** using the Multimeters. To place a Multimeter, click the *LEFT* mouse button on the **Multimeter** icon in the Instruments toolbar:

Place two Multimeters in your circuit as shown:

We will use **XMM1** to measure the voltage across **R2**, and use **XMM2** to measure the current through **L1**. Wire the meters as shown in the two screen captures below:

To set up and view meter results, we must open the meter windows by double-clicking on the meters. Double-click on each meter to open the windows. Arrange the windows as shown:

We must now set both meters to read AC waveforms, and set **XMM2** to read current and **XMM1** to read voltage. For **XMM2**, click on the [~] button and the [A] button to make the meter read AC current. For **XMM2**, click on the [~] button and the [V] button to make the meter read AC volts. When you have set up the meters correctly, click on the Run button [o|I] to restart the simulation. After a few moments, the meter windows will display their results:

We see that the voltage across R2 is equal to about 496 mV and the current through L1 is approximately 49.6 mA. We can use this method to find the magnitude of any voltage or current in the circuit. Remember that the values displayed are RMS values. The input was 1 volt RMS. If you want peak values, multiply all of the values by the square root of two.

EXERCISE 5-1: Find the RMS value of the voltages at nodes 1, 2, and 3 at a frequency of 1 kHz:

SOLUTION: Add three Multimeters or three voltage indicators and run the simulation:

EXERCISE 5-2: Find the magnitude of the voltage at node 5 and the current through **R1** at a frequency of 60 Hz:

HINT: **V2** is a current-controlled voltage source. The voltage of **V2** is **8** times the current through **R2**.

SOLUTION: Add Multimeters or voltage and current indicators. Note that when I added the current indicator to the circuit above, the node that was labeled **5** in the circuit above was renamed to **6** in the circuit below.

5.A.2. Magnitude and Phase Measurements – AC Analysis

We will now do the same analyses as in section 5.A.1 using the SPICE AC Analysis and we will also display the phase as well as the magnitude. We will start with the circuit below:

In order to use the AC Analysis, we must specify the properties of V1 for the sweep. The text displayed on the screen for V1 is only used with the virtual lab simulation and the instruments. There is another set of properties when the source is used for the AC Analysis. Double-click the **LEFT** mouse button on the V1 graphic:

Click on the **Analysis Setup** tab:

The dialog box states that this source will be used in the AC Frequency Analysis (the AC Analysis), the magnitude of the source is 1 volt, and the phase of the source is 0 degrees. The magnitude of the source can be interpreted as an RMS value or as a peak value. If we interpret the magnitude as an RMS value, then all voltages and currents we measure are RMS values. If we interpret the magnitude as a peak amplitude, then all measurements will be peak values. Our previous example used RMS values, so we shall interpret the 1 volt magnitude as 1 V RMS. The dialog box is already set up correctly, so click on the **OK** button to return to the schematic.

We would like to measure the current through **L1** and the voltage across **R2**. To measure current we will use a current monitor. As in section 4.A on page 174, we will use a current-controlled voltage source. Click the *LEFT* mouse button on the **Sources** button to obtain the **Sources** button bar:

Select the **Current_Controlled_Voltage_Source** button and place the element in your circuit:

Wire the source as shown below:

I have turned on the display of node names so that we can easily display node voltages. Note that the resistor voltage is the voltage at node **VR** relative to ground. I have also renamed the node between the current monitor and **R3** as **IL1**. To rename a node, double-click the *LEFT* mouse button on the wire. Note that resistor **R3** and a ground were added to the monitor circuit. The resistor was added so that there will be two elements connected to node **IL1**. The ground is necessary for the SPICE analysis because we cannot have floating nodes. The gain of the current monitor is **1 Ohm**, so **V2** is equal to **1 Ohm** times the current through **L1**, or the output of the monitor is equal to 1 volt per amp of measured current. Note that the unit of volts per amp (volts/amps) is equal to ohms.

We can now set up the AC Analysis. Select **Simulate**, **Analyses**, and then **AC Analysis** from the Multisim menus:

We want to analyze the circuit at a single frequency of 200 Hz. Fill in the dialog box as shown below:

We have specified a **Linear Sweep type** with total **Number of points** equal to 1. The **Start frequency** and **Stop frequency** are set to **200 Hz**. Thus, this dialog box specifies a single frequency at 200 Hz. We do not need to specify an option for the **Vertical scale** because we will not be looking at the results graphically.

Next, we need to specify the output variables we would like to observe. Click the *LEFT* mouse button on the **Output variables** tab:

Presently, no variables are selected, which means that we will not be able to display any information. We are interested in the voltages at nodes VR and IL1. Click on the text **vr** to select it:

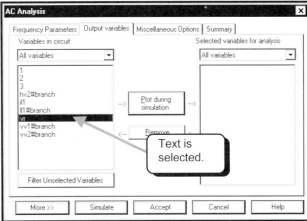

Click on the **Plot during simulation** button:

The voltage at node **vr** will now be plotted during the simulation and the data for the node will be available to the Postprocessor. Repeat the procedure to add node **il1** to the list:

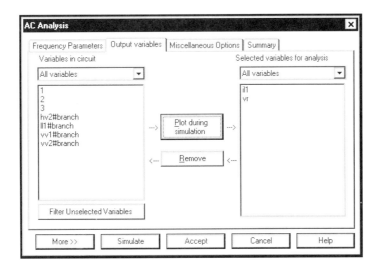

We are now ready to run the simulation, so click on the **Simulate** button:

By default, a plot is displayed. This plot is not too useful because our simulation contained only a single datum point. To view the results, we need to use the Postprocessor. Select **Simulate** and then **Postprocess** from the Multisim menus:

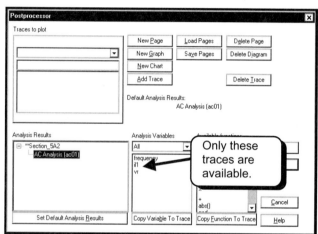

We see that the only data available to us is for the nodes we selected, **VR** and **IL1**.

To view the data as text, we need to create a new chart. First, click on the **New Page** button to create a new tab in the **Analysis Graphs** window.

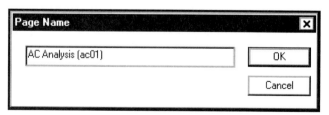

Enter a name for the new page, such as `Mag and Phase Data`, and click on the **OK** button. A new tab will be created:

EWB Multisim   Magnitude and Phase Simulations   223

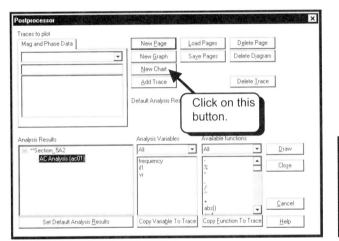

Next, click on the **New Chart** button:

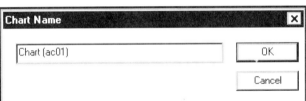

Enter a name for the chart, such as `AC Mag and Phase Text Output`, and click on the **OK** button:

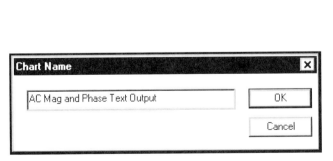

The next thing we need to do is to specify the data to be displayed in the chart. We will first display the frequency at which the analysis was run. Click the *LEFT* mouse button on the text **frequency** to select the item:

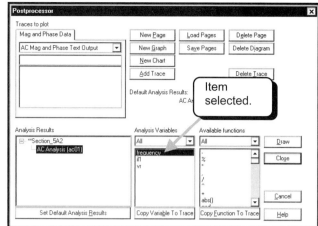

Click on the **Copy Variable To Trace** button:

Next, click on the **Add Trace** button:

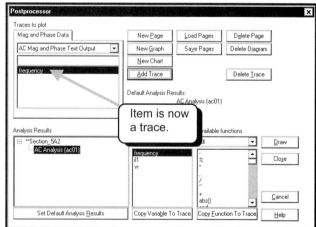

Next, we need to add the magnitude of the voltage at node **vr** to the list. We will add this trace in a different way. Click the *LEFT* mouse button as shown below:

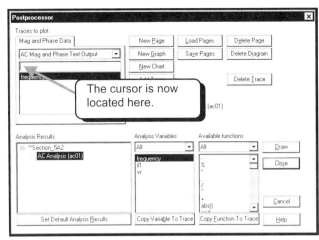

With the cursor located as shown, type the text **vm(vr)**. Multisim is case-insensitive, so you can use any combination of upper- or lowercase letters to create your traces.

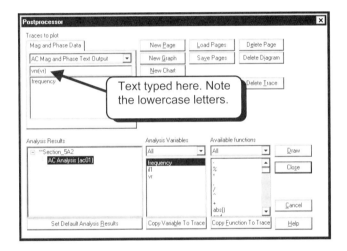

The **vm** function specifies that we would like to display the magnitude of the voltage at the selected node. In this case, we will display the magnitude of the voltage at node **vr** relative to ground. Click on the **Add Trace** button:

The magnitude of the voltage at node **vr** will now be displayed in the chart.

The function VP displays the phase of the voltage at a node in radians. Add the following traces: **vp(vr)*180/pi**, **vm(il1)**, and **vp(il1)*180/pi**:

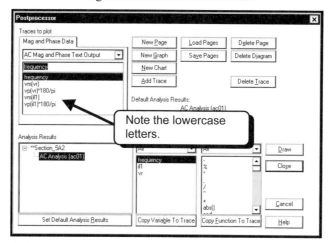

Remember that the voltage at node il1 is numerically equal to the current through L1. The above dialog box specifies that we will display the magnitude and phase of the voltage at node vr, the magnitude and phase of the current through L1, and the frequency.

We are now done adding traces. Click on the **Draw** button to create the chart:

Close the **Simulation Error Log/Audit Trail** window and close the **Postprocessor** window. Bring the **Analysis Graphs** window to the top:

The results show that the magnitude of the voltage at node **vr** is **496** mV with a phase angle of **-69.3**5 degrees ($0.496 \angle -69.35°$ volts). The current through L1 is **49.6** mA with a phase angle of **-69.3**5 degrees ($0.0496 \angle -69.35°$ amps).

EXERCISE 5-3: Find the magnitude and phase of the voltages at nodes 1, 2, and 3 at a frequency of 1 kHz using the AC Analysis.

SOLUTION: In the **Analysis Setup** for the two sources, specify **I1** as a 2 A source and **V1** as a 12-volt source. These values duplicate the RMS values used in **EXERCISE 5-1**.

Set up the **AC Analysis** dialog boxes as shown:

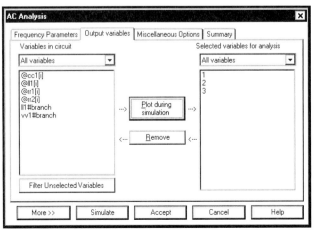

The magnitude and phase results are:

EXERCISE 5-4: Find the magnitude and phase of the voltage at node 5 and the current through **R1** at a frequency of 60 Hz using the AC Analysis:

SOLUTION: Add a current sensor as shown:

Magnitude and Phase Simulations

In the **Analysis Setup** for the two sources, specify **I1** as a 5.66 A source and **V1** as a 20-volt source. These values duplicate the RMS values used in EXERCISE 5-2.

Set up the **AC Analysis** dialog boxes as shown:

The magnitude and phase results are:

5.B. Bode Plots

Bode plots are plots of magnitude and phase versus frequency. Since we are usually interested in the magnitude of the gain, an AC 1-volt source will be used. All AC analyses assume a linear network: If the output for a 1 V source is 3 V, the output for a 10 V source will be 30 V. Since gain is the ratio of an output to the source, and since the networks are linear, the magnitude of the input does not matter. For convenience we will use a 1-volt magnitude source. Bode phase plots will give us the phase of any voltage or current relative to the phase of the source. For simplicity we will set the phase of the source to zero.

As with most simulations in Multisim, we can simulate the circuit using a virtual laboratory with the Bode Plotter instrument, or we can use the standard SPICE AC Analysis. The Bode Plotter can only be used for Bode plots while the AC Analysis is a more general analysis and can be used for other applications in addition to Bode plots. We will show both methods in this section.

5.B.1. Bode Plots Using the Bode Plotter Instrument

We will first illustrate how to use the Bode Plotter instrument. This method gives us the quickest way of observing the frequency response of a network. The limitation is that we can only view a single node in the circuit. If you have a large circuit and want to look at the response of several nodes, you should use the AC Analysis.

Wire the low-pass filter shown:

At low frequencies, the capacitor is an open circuit and **Vout** should equal **Vin**. At high frequencies the capacitor becomes a short, and the gain goes toward zero. The 3 dB frequency of the circuit is $\omega = 1/R_1C_1 = 1{,}000$ rad/s $\cong 159$ Hz.

Next, we need to place the Bode Plotter instrument in the circuit. The Bode Plotter button is located in the Instruments toolbar:

Place the Bode Plotter and connect it to your circuit as shown:

We can either run the simulation now, or modify the parameters for the Bode plot. To illustrate how to change the settings, we will modify the parameters. Double-click the **LEFT** mouse button on the Bode Plotter instrument to open the instrument window:

If we look at the instrument window a little closer, we see the following settings:

The settings specify that the instrument window will display the **Magnitude** plot, the y-axis will have a range of **0 dB** to **-200 dB**, and the x-axis will have a range of **1 mHz** (10^{-3} Hz) to **1 GHz** (10^9 Hz). The x-axis specifies the range of frequencies that the frequency will be swept. The y-axis is just an initial guess as to what the range of output gain might be. Since I know that the pole for this circuit is at 159 Hz, I will change the x-axis to sweep frequencies from 100 mHz to 10 MHz:

The Bode Plotter does not give us a very big window to view results, so we will not worry too much about the settings for the y-axis.

Click on the **Run** button to simulate the circuit. The Bode Plotter will trace out the Bode magnitude plot, which is the ratio of the magnitude of the voltage measured by the **out** terminals divided by the magnitude of the voltage measured by the **in** terminals:

EWB Multisim   Magnitude and Phase Simulations   233

We see that the y-axis range was not a good guess because the plot only traverses about one-half the range of the y-axis. Also, the Bode Plotter does not number the axes. To get a better view of the results, we will use the Grapher.

Before we use the Grapher, we must close the Bode Plotter window. Click the **LEFT** mouse button on the ☒ in the upper right corner of the Bode Plotter window to close the window:

We can now view the results with the Grapher. Select **View** and then **Show Grapher** from the Multisim menus:

We can now modify the plot using the techniques covered in section 2.F. To change the properties of a plot, click the **RIGHT** mouse button on the plot you wish to change. For example, click the **RIGHT** mouse button on the lower plot and then select **Properties**:

Use this dialog box to turn on the grid, set the title to "Low Pass Filter – Bode Phase Plot," change the left axis range to –90 to 0 degrees, and specify a trace pen size of 2. When done, click on the **OK** button:

To modify the top plot, click the **RIGHT** mouse button on the top plot and select **Properties**. Make the same changes as we did for the phase plot except change the title to "Low Pass Filter – Bode Magnitude Plot," and change the y-axis range to –100 to 0 dB:

We can also use the cursors on the top plot to find the –3 dB frequency. To display the cursors, click the **RIGHT** mouse button on the top plot and select **Toggle Cursors**:

Place cursor 1 (red) so that it reads a **y1** value close to 0 dB. Place cursor 2 (blue) so that the cursor display box reads a **dy** value close to –3 dB:

The x coordinate of cursor 2 (**x2**) specifies the frequency coordinate of cursor 2. This value is the –3 dB frequency and is approximately 162 Hz in this example. Our calculated value was 159 Hz. The values differ slightly because we could not hit the –3 dB point exactly with the cursors.

5.B.2. Bode Plots Using the AC Analysis

We will now simulate the same circuit using the SPICE AC Analysis. Wire the circuit below, or use the circuit used in the previous example. Note that we have removed the Bode Plotter instrument because it is unnecessary when running the AC Analysis.

We first need to set up voltage source V1 for the sweep. To do this, double-click the *LEFT* mouse button on the graphic:

Select the **Analysis Setup** tab:

We are not running any of the distortion analyses, so we can ignore the distortion options of this dialog box. We are interested in the top set of options. These options specify that this source will be used in the **AC Frequency Analysis** (also referred to in other SPICE engines as AC Analysis), and that the magnitude of the source will be 1 volt and its phase will be zero degrees. In phasor notation, V1 is $5\angle 0°$.

Since the gain of this circuit is the ratio of the voltage at node Vout to the voltage at node Vin, using a 1-volt source is as good a choice as any, and it also makes the ratio easier for humans to calculate. We do not need to change any of the parameters for this run. However, now that we have seen these properties, we know where to find them when we do need to change them. Click on the **OK** button to return to the schematic.

Next, we need to set up the AC Analysis and then run the analysis. Select **Simulate**, **Analyses**, and then **AC Analysis** from the Multisim menus:

This first tab allows us to specify the range of frequencies for the sweep, the sweep type, and the number of points in the sweep. A decade is a factor of 10 in frequency. For example, 1 Hz to 10 Hz is a decade, 10 Hz to 100 Hz is a decade, 100 Hz to 1000 Hz is a decade, and so on. When you specify a sweep in **Decade**s, you also specify the number of points per decade. For example, with the current settings, there will be 10 simulation points between 1 and 10 Hz, 10 simulation points between 10 Hz and 100 Hz, and so on.

You can also specify sweep types of Octave and Linear. An octave is a factor of two in frequency, 1 Hz to 2 Hz, 2 Hz to 4 Hz, and so on. For an octave sweep type, you will specify the number of points per octave. For a linear sweep type, you will specify the total number of points in the sweep. The linear sweep is useful for frequency ranges where the starting frequency is of the same order of magnitude as the ending frequency, say 1 kHz to 5 kHz. A decade sweep is useful for wide frequency ranges that span several orders of magnitudes, say 1 Hz to 1 MHz.

We will set up this sweep with the parameters we specified in the Bode Plotter setup on page 232:

The dialog box is set to sweep frequencies from **0.1 Hz** (100 mHz) to **10 MHz** in decades, with **100** points per decade.

The last thing we need to do is to set the **Vertical scale** for the magnitude plot. Click the *LEFT* mouse button on the down triangle ▼ as shown:

The **Linear** option specifies that the y-axis for the magnitude plot is to be a linear scale while the **Logarithmic** option specifies the y-axis as a log base 10 scale. With a log base 10 scale, factors of 10 are spaced the same distance apart. That is, the spacing between the number 1 and 10 is the same as the spacing between 10 and 100, which is the same as the spacing between 100 and 1000, and so on. The **Octave** option specifies that the y-axis is to be logarithmic in factors of 2. With a log base 2 scale, factors of 2 are spaced the same distance apart. That is, the spacing between the numbers 1 and 2 is the same as the spacing between 2 and 4, which is the same as the spacing between 4 and 8, and so on. The **Decibel** option specifies that the magnitude plot should be plotted in decibels ($20\log 10(v_{out}/v_{in})$). In a standard Bode magnitude plot, the magnitude in decibels is plotted versus frequency, so we will select the **Decibel** option:

238 Magnitude and Phase Simulations EWB Multisim

Next, we must specify the output variables for the simulation. Select the **Output variables** tab:

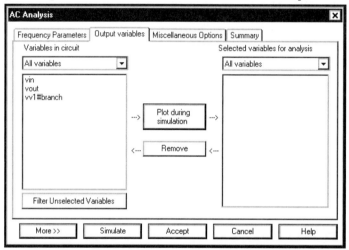

Click on the text **vout** to select it, and then click on the **Plot during simulation** button to add the variable to the right window pane:

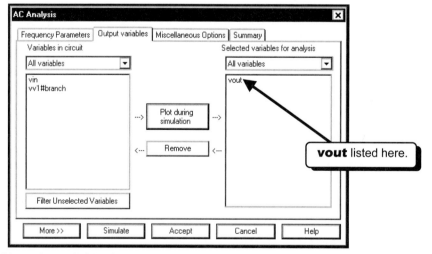

The data from node **vout** will now be available after the simulation.

We are now ready to run the simulation, so click on the **Simulate** button:

We see that the magnitude and phase plots of the voltage at node **vout** are automatically displayed. These plots are nearly the same as we obtained from the Bode Plotter shown on page 235. We can use the techniques shown on pages 233 to 235 to obtain the same plot as shown on page 235:

The cursors show a –3 dB frequency of **162.1810** Hz. This is slightly different than the value from the previous section. This is because we specified 100 points per decade in this simulation setup, while the number of points per decade for the Bode Plotter may be different. In either case, both –3 dB frequencies are slightly off the calculated value because we could not hit the –3 dB frequency exactly with the cursors. To come up with a closer value, we must specify more points per decade in our AC Analysis setup.

EXERCISE 5-5: Plot the Bode phase and magnitude plots for frequencies from 1 Hz to 1 MHz. Use the cursors to find the frequencies of the poles and zeros.

ANALYTIC SOLUTION: At low frequencies, the capacitor is an open circuit and the voltage at **Vout** can be obtained by the voltage divider of **R3** and **R4**:

$$\frac{V_{out}}{V_{in}} = \frac{R_4}{R_3 + R_4} = 0.0099 = -40 \text{ dB}$$

At high frequencies, the capacitor is a short circuit and **Vout** = **Vin** so

$$\frac{V_{out}}{V_{in}} = 1 = 0 \text{ dB}$$

Thus, we should expect the magnitude plot to start at –40 dB and finish at 0 dB. The next question is, where are the poles and zeros? Let

$$Z = Z_c \| R_3 = \frac{(1/j\omega C_2)R_3}{(1/j\omega C_2) + R_3} = \frac{R_3}{1 + j\omega R_3 C_2}$$

Substituting Z for $Z_c \| R_3$ yields the equivalent circuit:

An expression for the gain can now be obtained from the voltage divider of **R4** and **Z**:

$$\frac{V_{out}}{V_{in}} = \frac{R_4}{Z + R_4} = \frac{R_4}{\left(\dfrac{R_3}{1 + j\omega R_3 C_2}\right) + R_4}$$

Multiplying the numerator and denominator by $1 + j\omega R_3 C_2$ gives

$$\frac{V_{out}}{V_{in}} = \frac{R_4(1 + j\omega R_3 C_2)}{R_3 + R_4 + j\omega C_2 R_3 R_4}$$

We see that the zero is at $1 = \omega R_3 C_2$, or

$$\omega_z = \frac{1}{R_3 C_2} = \frac{1}{(1500 \, \Omega)(1 \, \mu F)} = 666 \text{ rad/s} = 106 \text{ Hz}$$

The pole is at $R_3 + R_4 = \omega C_2 R_3 R_4$, or

$$\omega_p = \frac{R_3 + R_4}{C_2 R_3 R_4} = \frac{1}{C_2 \left(\dfrac{R_3 R_4}{R_3 + R_4}\right)} = \frac{1}{C_2 (R_3 \| R_4)}$$

$$= \frac{1}{(1 \, \mu F)(1500 \, \Omega \| 15 \, \Omega)} = 66.6 \text{ krad/sec} = 10.6 \text{ kHz}$$

SOLUTION: Use the Bode Plotter and specify frequencies from 1 Hz to 1 MHz. Plot the results with the Grapher and use the cursors to locate the numerical values:

The cursors above show the location of the zero to be at **105.6728** Hz, close to the calculated value of 106 Hz. Next, we use the cursors to find the location of the pole:

The cursors show that the location of the pole is at **10.7152 k**Hz, close to the calculated value of 10.6 kHz.

EXERCISE 5-6: Plot the Bode phase and magnitude plots for frequencies from 1 Hz to 1 MHz. Use the cursors to find the frequencies of the poles and zeros.

ANALYTIC SOLUTION: At low frequencies, the capacitor is an open circuit and the voltage at **Vout** is zero. As the frequency is increased, the impedance of the capacitor will decrease and the voltage at **Vout** will increase. Thus, at low frequencies we should expect the gain in decibels to be a large negative number and to increase at a rate of 20 dB/decade. At high frequencies, the capacitor is a short circuit and **Vout** can be obtained from the voltage divider of R1 and R2:

$$\frac{V_{out}}{V_{in}} = \frac{R_2}{R_2 + R_1} = 0.5 = -6 \text{ dB}$$

Thus, we should expect the magnitude plot to start at a negative value in decibels and finish at –6 dB. The next question is, where are the poles and zeros? The gain of the circuit can be obtained from the voltage divider of **R1**, **R2**, and **C1**:

$$\frac{V_{out}}{V_{in}} = \frac{R_2}{R_1 + R_2 + Z_c} = \frac{R_2}{R_1 + R_2 + \left(\frac{1}{j\omega C_1}\right)} = \frac{j\omega C_1 R_2}{1 + j\omega C_1 (R_1 + R_2)}$$

We see that there is a zero at ω = 0 rad/sec and a pole at 1 = ωC$_1$(R$_1$ + R$_2$), or

$$\omega_p = \frac{1}{C_1(R_1 + R_2)} = \frac{1}{(100 \text{ nF})(1000 \text{ }\Omega + 1000 \text{ }\Omega)} = 5000 \text{ rad/s} = 795.8 \text{ Hz}$$

SOLUTION: Set up an AC Analysis to sweep the frequency from 1 Hz to 1 MHz at 100 points per decade.

Plot **vout** and use the cursors to locate the –3 dB point.

The cursors show the location of the pole to be at **794.3282** Hz. This is close to the calculated value of 795.8 Hz.

5.C. Amplifier Gain Analysis

One of the most important applications of the AC Analysis and the Bode Plotter instrument is to see the frequency response of an amplifier. If an AC Analysis or Bode Plotter is used on a circuit with a transistor, the DC bias point is calculated and the transistor is replaced by a small-signal model around the bias point. The frequency response is then measured using the linearized model of the transistor. The AC Analysis and the Bode Plotter instrument can only be used to find the small-signal gain and frequency response. Voltage swing, clipping, and saturation information must be obtained from the transient simulation, or by using the operating point information.

For this example, we could use either the AC Analysis or the Bode Plotter instrument. Both should yield the same results. For this example, we will use the Bode Plotter instrument. Wire the amplifier circuit as shown:

The 2N3904 transistor is located in the **Transistors** toolbar. To access this toolbar, click on the **Transistors** button as shown:

Click on the NPN_BJT button to view the list of NPN transistors available:

You can now select a 2N3904 and place it in your circuit.

We would like to see the gain of this circuit for frequencies from 1 Hz to 100 MHz. Place the Bode Plotter instrument in your schematic and wire it to nodes **IN** and **Out**:

We must now change the parameters for the Bode plot. Double-click on the Bode Plotter instrument to bring up its window:

The instrument is set to sweep frequency from **1 mHz** to **1 GHz**. For our Bode plot, we would like to simulate the circuit from 1 Hz to 100 MHz, so modify the dialog box as shown:

We are now ready to run the simulation, so click on the **Run** button to start the simulation. The Bode magnitude plot for this amplifier will be displayed on the Bode Plotter instrument screen:

Where is the plot? It turns out that the default range for the **Vertical** axis (y-axis) is from –200 dB to 0 dB. The gain of this amplifier at midband is about 45 dB. This value is out of range for the **Vertical** scale. Change the **Vertical** scale to range from 15 dB to 55 dB. As you change the scale, the Bode plot will appear within the **Bode Plotter** instrument window:

If we move the cursor to the center of the plot, we measure a gain of 45.716 dB, close to the expected value:

We can use the cursor to locate the upper and lower –3 dB frequencies of this plot. The midband gain is 45.7 dB, so the –3 dB frequencies will be located at the frequencies where the gain is 3 dB less than the midband gain, or 42.7 dB. We can locate these two points with the cursor:

We see that the lower –3 dB frequency occurs at **64.565 Hz** and the upper –3 dB frequency occurs at **6.457 MHz.**

EXERCISE 5-7: Find the mid-band gain and upper and lower –3 dB frequencies of the common-base amplifier shown below.

SOLUTION: Add the Bode Plotter instrument and simulate the circuit for frequencies from 10 mHz to 100 MHz:

5.D. Operational Amplifier Gain

In this section we will use Multisim to determine the gain and bandwidth of an operational amplifier circuit with negative feedback. Since we used the Bode Plotter in the last section, we will use an AC Analysis for this example. We will find the gain and upper –3 dB frequency for the circuit below:

This circuit contains a few unfamiliar parts, so we will show how to draw the circuit here. First, we will place the LM124J OPAMP. To place this part, click the *LEFT* mouse button on the **Analog** button as shown:

Select the **OPAMP** button:

Locate and select the **LM124J** in the **Component Name List**:

Click on the **OK** button to place the part:

The LM124 is a quad OPAMP package. That is, a single package contains four OPAMPs. If you look at a datasheet for a LM124, you will see that each package contains four OPAMPs designated A, B, C, and D. This menu allows us to select a specific OPAMP in the quad package. For simulations, the OPAMP we select is not important, so select one of the OPAMPs and place the part in your circuit:

The next thing we will do is place the DC supplies in the circuit. We could do this using standard DC supplies as shown below (don't do this):

This method works fine, but it does clutter up the schematic a bit. Instead of using these DC supplies, we will use a part called Vcc. Close the analog parts toolbar and then click on the **Sources** button:

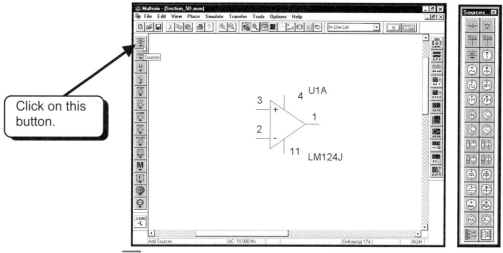

Click on the **Vcc** button to select the part:

Place two copies of the part as shown:

The first thing we will do is change the properties of the top **VCC** part. Double-click the *LEFT* mouse button on the graphic as shown below:

This tab sets the voltage of the node at **5** volts. Change the value to **15** volts:

Next, we want to change how the part displays its information. Select the **Display** tab:

This part uses the Multisim defaults for specifying what information to display about a part. Deselect the option as shown:

For our part, two values are displayed on the schematic. The **15V** is the value of the source and its display is enabled by selecting the **Show values** option, which is enabled. The text **Vcc** is the reference ID for this part and its display is enabled by the **Show reference ID** option, which is also selected. We do not want to display the text Vcc, so disable the **Show reference ID** option:

Click on the **OK** button to accept the changes and return to the schematic:

We see that the text Vcc is no longer displayed and that the value of the part has changed to **15V**.

Next, we need to change the parameters of the lower **Vcc** part. Change the value to **−15 V** and disable the **Show reference ID** option as shown below:

The last thing we need to change is the reference ID for this part. If we do not change the reference ID, both nodes will be labeled as Vcc and will be electrically connected together. This would cause problems, so we will change the reference ID. Select the **Label** tab:

Change the **Reference ID** to **Vee**:

Click on the **OK** button to accept the changes:

Connect the two parts to your circuit:

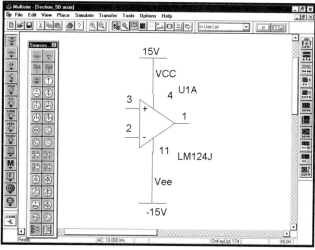

On my screen capture the text **VCC** and **Vee** is displayed. This is because I have selected the global option to display node names on my circuit so that we can easily label nodes and see their labels. You may not have this option selected, so you may not see the **VCC** and **Vee** labels. If you cannot see the labels, you can double-click on a wire to see its properties. For example, if I double-click on the wire as shown below, you will see information about the wire:

We see that the node is named **Vee**. If your nodes are not named differently (VCC and Vee) you need to double-click on the two parts again and change their reference IDs to make the parts different. Click on the **OK** button to close the window. The remainder of the circuit is similar to ones we have done before, so complete the circuit as shown:

Magnitude and Phase Simulations

We would like to find the gain of this amplifier, V_o/V_{in}. Since the magnitude of V_{in} is one volt, the magnitude of V_o is the gain. If the gain of the amplifier is 200, then the magnitude of the output will be 200 V. This may seem a little unreasonable since the DC supplies are ±15 V. When an AC Analysis is performed, the operating points of all non-linear parts are found, and the parts are replaced by their linear models. Since the models are linear, the voltages and currents in the circuit are not limited by the DC supplies. For a linear circuit, if an input of 1 V produces an output of 10 V, then an input of 1,000,000 V will produce an output of 10,000,000 V. Thus, for gain purposes the magnitude of the input does not matter and a 1 V input is chosen for convenience. For an AC simulation a magnitude of 200 V is not unreasonable. A magnitude of 200 V is not physically possible for our OPAMP, but it is a valid number when using the AC Sweep to calculate the gain. If you wish to observe the maximum voltage output of a circuit, you must run one of the Time Domain Analyses. An example of the voltage swing of a transistor amplifier is shown in Section 6.H on page 368. The same techniques could be used to find the voltage swing limits of this OPAMP circuit.

The next thing we need to do is set up the AC source to be used in the AC Analysis. Double-click the **LEFT** mouse button on the sinusoidal source and then select the **Analysis Setup** tab:

These settings specify that the source we selected will be used in the AC Analysis and that it will have a magnitude of 1 V and a phase of zero degrees. This is the default setting, but it is a good idea to check this setting before running an AC Analysis. We are not going to be running a distortion analysis, so the other settings under this tab are not relevant. Click on the **OK** button to return to the schematic.

Next, we need to set up the simulation. Select **Simulate**, **Analyses**, and then **AC Analysis** from the Multisim menus:

We would like to simulate this circuit from 1 Hz to 10 MHz. Since our frequency range spans several orders of magnitude in frequency (several decades of frequency), we will use the **Decade Sweep type**. To get a detailed plot, we will choose 100 **points per decade**. For a Bode plot, we would like the y-axis to be plotted in decibels, so we will select **Decibel** for the **Vertical scale**:

The last thing we must do is specify the variable we would like to plot. Select the **Output variables** tab and specify **vo** as an output variable to plot during the simulation:

We are now ready to run the simulation, so click on the **Simulate** button:

EWB Multisim   Magnitude and Phase Simulations   257

We can now use the cursors to find the midband gain and the upper –3 dB frequency. Click the **RIGHT** mouse button on the upper plot and then select **Toggle Cursors**:

Use cursor one to find the low-frequency gain and use cursor two to find the frequency where the gain is down by 3 dB from the low-frequency gain:

Coordinate **y1** shows us that the gain at cursor one is **29.5396** dB, or a gain of about 30. **dy** is the difference between the y-coordinates of cursor 1 and cursor 2. Thus, when **dy** = –3, we know that the gain at cursor two is 3 dB less than the gain at cursor one. Thus, when **dy** = –3, we know that cursor 2 is placed at the –3dB frequency. Coordinate **x2** shows us that the 3 dB frequency is **31.6228 k**Hz.

As a side note, we see that the midband gain of this amplifier is 30, and the –3dB frequency is 31.622 kHz. The gain bandwidth product of this amplifier is:

$$GBW = (30)(31.6228 \text{ kHz}) = 948,684 \text{ Hz}$$

This agrees well with the GBW specification of the LM124, which is specified as 1 MHz.

EXERCISE 5-8: Find the midband gain and upper –3 dB frequency for the amplifier shown below, and find the gain-bandwidth product (GBW) from the simulated results.

SOLUTION:

The midband gain is **38.0544** dB, which corresponds to a gain of 79.932. The upper –3 dB frequency is **11.7490 k**Hz. The gain-bandwidth product is

$$GBW = (79.932)(11.7490 \text{ kHz}) = 939121 \text{ Hz}$$

This value is very close to the value from the previous example. This is expected since the GBW product should be approximately constant for all circuits made from the same OPAMP.

5.E. Parameter Sweep — OPAMP Gain Bandwidth

In this section we will use Multisim to determine the bandwidth of an OPAMP circuit with varying amounts of negative feedback. For an OPAMP circuit, the closed-loop gain times the bandwidth is approximately constant. To observe this property, we will run a simulation that creates a Bode plot for several different closed-loop gains. We will use the circuit below:

This circuit uses a **LM124J** OPAMP macro model. All of the real OPAMP models include bias currents, offset voltages, slew rate limitations, and frequency limitations. Also note that real OPAMP models require DC supplies. The models will not work without the supplies.

We would like to see the frequency response of the OPAMP circuit for different values of the gain, so we must set up a Parameter Sweep and then specify the analysis type to be an AC Analysis. Select **Simulate**, **Analyses**, and then **Parameter Sweep**:

The **Sweep Parameter** field can be either a **Device Parameter** or a model parameter. A **Device Parameter** is something like the resistance of a resistor, the inductance of an inductor, the temperature coefficient of a capacitor, and so on. A model parameter would be a parameter in a diode or transistor model. Examples would be the breakdown voltage of a Zener diode (model parameter BV) or the forward current gain of a bipolar junction transistor (model parameter BF).

We would like to observe the frequency response of this OPAMP circuit for different gains. The gain of this circuit is $1 + \dfrac{R_1}{R_3}$, so we can change the gain by changing the value of R_1 or R_3. To change the gain, we will change the value of R_1, so we will use the Parameter Sweep to change the value of the resistance of R_1. Resistance of a resistor is a **Device Parameter**, so the **Sweep Parameter** should be set to **Device Parameter** as shown above. The **Device** field lists all of the available devices we can specify such as diodes, transistors, capacitors, and resistors. Select the **Resistor Device**.

Once you select the **Resistor Device**, all available resistors in your circuit will be listed in the **Name** field:

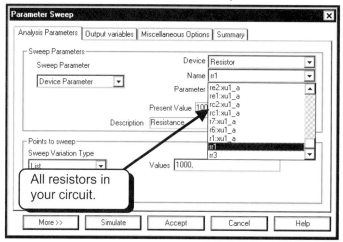

The pull-down menu lists many more resistors than we see on the schematic. This is because this list includes all of the resistors used in the OPAMP subcircuit. A prefix of r is added to all resistors, so **rr1** is resistor **R1** in our circuit. Select this item:

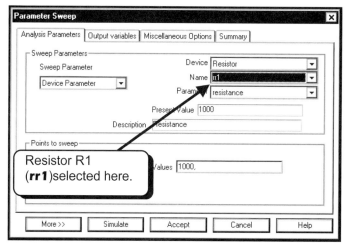

Next, we need to specify what parameter of the resistor we want to sweep. The parameter field lists all of the items we can sweep for a resistor:

We see that we can sweep the temperature at which this resistor operates (**temp**), the length and width of the resistor (**l** and **w** – used for integrated circuits and film resistors), and other properties as well. We want to change the resistance, so select **resistance**. Note that the **Description** field describes the function of the selected **Parameter**:

Next, we need to specify the values we want R1 to have. We will use the **List Sweep Variation Type** and specify the values for R1 to be 1000, 10000, 100000, 1000000. Values are separated by commas:

Note that we could have specified the same values for R1 using a Decade **Sweep Variation Type**.

We have now finished the setup for the **Parameter Sweep**. The next thing we must do is specify that this **Parameter Sweep** should run in conjunction with an AC Analysis. Click on the **More** button:

The default **Analysis to sweep** is a **Transient analysis**. To find the frequency response we need to run an AC Analysis. Select the **AC analysis** and also specify that we want to **Group all traces on one plot**. Selecting this option will allow us to view all of the results from the **Parameter Sweep** on a single plot. Select options as shown below. Make sure that your dialog box matches all of the options selected below:

Next, we must set up the parameters for the AC Analysis. Select the **Edit Analysis** button as shown below and set up the parameters for the **AC analysis** as shown. This setup is the same as we have done for analyses in section 5.D:

Click on the **Accept** button to return to the **Parameter Sweep** dialog box.

EWB Multisim   Magnitude and Phase Simulations   263

We have specified the following simulation. First, R1 will be set to 1,000 Ω and then the AC Analysis will sweep from 1 Hz to 10 MHz. Next, R1 will be set to 10,000 Ω and then the AC Analysis will sweep from 1 Hz to 10 MHz. Next, R1 will be set to 100,000 Ω and then the AC Analysis will sweep from 1 Hz to 10 MHz. Finally, R1 will be set to 1,000,000 Ω and then the AC Analysis will sweep from 1 Hz to 10 MHz. We will look at the results from all of the AC Analyses on a single plot.

The last thing we need to do is to specify the data we wish to plot. Select the **Output variables** tab and specify that **vo** should be plotted during the simulation:

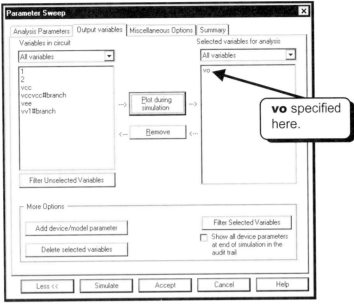

We are now ready to run the simulation, so click on the **Simulate** button:

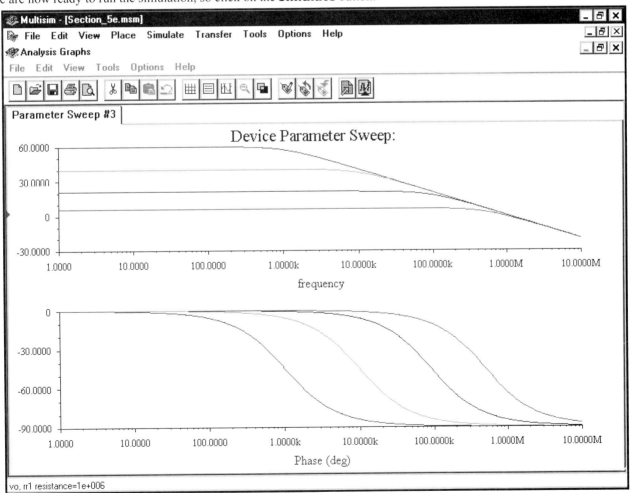

The top plot shows us the gain in decibels versus frequency. We see that higher gain rolls off at lower frequency. Using the cursors we find that for a gain of 60 dB (1000) the upper –3 dB frequency is 1 kHz. This corresponds to a gain bandwidth of 1 MHz (1000 times 1000). For a gain of 40 dB (100) the upper –3 dB frequency is 10 kHz. This also corresponds to a gain bandwidth of 1 MHz (100 times 10,000). For a gain of 20 dB (10) the upper –3 dB frequency is 91.2 kHz, corresponding to a gain bandwidth of 912 kHz (10 times 91,200). For a gain of 3 dB (2) the upper –3 dB frequency is 490 kHz, corresponding to a gain bandwidth of 980 kHz (2 times 490,000).

We see that for all of these gains, the gain-bandwidth product is relatively constant. This means that for high-gain OPAMP circuits the gain will roll-off at lower frequencies, and vice versa. This also means that we can trade gain for bandwidth for any particular OPAMP circuit. If we need a larger or smaller bandwidth, we can change the gain to achieve that bandwidth.

EXERCISE 5-9: For the inverting OPAMP below, find the upper –3 dB frequency for mid-band gains of –1, –10, –100, and –1000. Show that, except for the gain of –1, the magnitude of the gain times the –3 dB bandwidth is relatively constant for all of the plots. This circuit with a gain of –1 is one of the few exceptions where the gain times bandwidth is not constant.

SOLUTION: Run the Parameter Sweep with the AC Analysis.

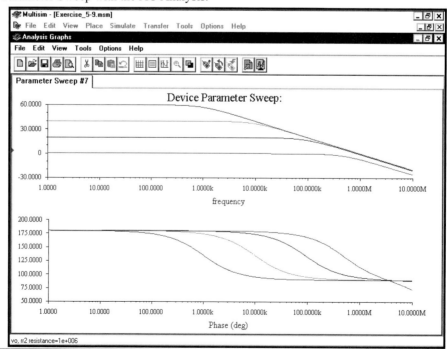

5.F. AC Power and Power Factor Correction

The AC power absorbed by a load and the power factor of the load can be easily measured using the Multisim Wattmeter. The Wattmeter can be used to measure both DC and AC power. In section 3.A.3, we demonstrated its use with the DC power absorbed by a resistor in a circuit with a DC source. Here we will demonstrate its use by finding the power absorbed by a complex load and the power factor of the load.

All linear loads can be reduced to an equivalent single impedance Z_{Load}, where Z_{Load} can be expressed as a magnitude and a phase, $Z_{Load} = Z_L \angle \theta_L$, or in rectangular coordinates as $Z_{Load} = A + jB$. In an AC circuit, the power absorbed by this load is

$$P_{Load} = \frac{V_M I_M}{2}\cos(\theta_L) = V_{RMS} I_{RMS} \cos(\theta_L)$$

where V_M and I_M are the peak values of the load voltage and current, and V_{RMS} and I_{RMS} are the RMS values of the load voltage and current. θ_L is the angle of the load and is also the angle between the load voltage and the load current. $\cos(\theta_L)$ is called the power factor (PF) and is a measurement of how inductive or capacitive a load appears. The Wattmeter displays the AC power absorbed by a load and the power factor.

As an example, we will calculate the power absorbed by the load below:

We will assume that this load is connected to the AC line where the voltage is 115 VRMS at 60 Hz. The impedance of this load at 60 Hz is

$$Z_{Load} = R_1 + j\omega L_1 = 3\Omega + j(2\pi 60\,\text{Hz})(0.004\,\text{H}) = (3 + j1.508)\Omega = (3.358\angle 26.687°)\Omega$$

If this load is connected to the 115 VRMS line, the power absorbed by this load is

$$P_{Load} = V_{RMS} I_{RMS} \cos(\theta_L) = V_{RMS}\left(\frac{V_{RMS}}{|Z_{Load}|}\right)\cos(\theta_L) = \frac{(115V)^2}{3.358\,\Omega}\cos(26.687°) = 3.519\text{ kW}$$

The power factor is

$$PF = \cos(\theta_L) = \cos(26.687°) = 0.893$$

We have now made a few simple calculations and we can compare them to results of Multisim simulations. We would like to measure the power absorbed by this load when connected to a 115 VRMS source as shown below:

The voltage displayed (**162.63**) is the peak value of the source. If we divide this peak value by $\sqrt{2}$, we will obtain the RMS value, which is 115 VRMS. Thus, we have connected our load to a 115 VRMS source. We would like to measure the power absorbed by the load when connected to this source using the Wattmeter. This meter measures the load current and voltage and displays the power absorbed and the power factor.

To place the Wattmeter in your circuit, click the **LEFT** mouse button on the Wattmeter button in the Instruments toolbar. The Wattmeter will become attached to your mouse pointer:

Place the Wattmeter in a convenient location in your circuit.

To measure the power absorbed by the load, we must measure the current through the load and the voltage across the load. Measuring current with the wattmeter is the same as measuring current with a Multimeter or with a current indicator; if current enters the positive terminal and leaves the negative terminal, the meter will interpret it as a positive current. Wire the meter as shown:

As shown, the meter measures the top terminal of the load as the positive voltage reference and the bottom terminal of the load as the negative reference. Current is measured as positive entering the top terminal of the load. Thus, the positive direction of current is entering the positive voltage terminal. Note that our equation for power, $P_{Load} = V_{RMS} I_{RMS} \cos(\theta_L)$, applies only if you follow the convention that the positive direction of current enters the positive voltage reference terminal.

One last thing to note in the circuit above is that we added a ground to the circuit. Although this is not necessary for the real circuit, Multisim and SPICE require all circuits to be grounded. If you do not ground your circuit, you will receive an error message.

Double-click the **LEFT** mouse button on the power meter to open its window:

We are now ready to run the simulation. Select **Simulate** and then **Run** from the Multisim menus, press the **F5** key, or click on the Simulate button to start the simulation.

We see that the load absorbs **3.517 kW** of power and that the power factor is 0.893. Both of these values agree with our calculated values.

5.F.1. Power Factor Correction

Now that we have found the power absorbed by the load and the power factor using calculations and simulation, we can look at correcting the power factor to unity. The basic idea behind power factor correction is that we can add an element in series or parallel with the load to cancel out the reactive portion of the load and make the load look pure real. A pure real load has a unity power factor. For the inductive load of the previous example, the easiest way to correct the power factor and make the load look pure real is to add a capacitor in parallel with the load.

Remember that admittances add in parallel. Since we are placing a capacitor in parallel with our load, the admittance of the capacitor will add to the admittance of the load. We must first calculate the admittance of our load:

$$Y_{Load} = \frac{1}{Z_{Load}} = \frac{1}{(3+j1.508)\Omega} = \frac{1}{(3.358\angle 26.687°)\Omega} = (0.298\angle -26.687)\text{S} = (0.266 - j0.134)\text{S}$$

The admittance of a capacitor is $Y_C = j\omega C$. When we place the capacitor in parallel with our load, the total admittance is

$$Y_{Total} = Y_{Load} + Y_C = (0.266 - j0.134) + j\omega C = 0.266 + j(\omega C - 0.134)$$

To make the load admittance pure real we need $\omega C - 0.134 = 0$, or $C = 0.134/\omega$. Our example is at a frequency of 60 Hz, or $\omega = 2\pi 60$. For this frequency, the needed capacitor is 354.8 µF.

To check our calculations and method, we will simulate the circuit of the previous example with the addition of this parallel capacitor. Add the capacitor as shown:

Select **Simulate** and then **Run** from the Multisim menus, press the **F5** key, or click on the Simulate button to start the simulation.

We see that the load absorbs the same amount of power as the previous simulation. This is the expected result because we placed the capacitor in parallel with the load, so the voltage across the load is the same for both simulations. If the voltage across the load is the same in both simulations, the power absorbed by the load must be the same. Capacitors and inductors do not absorb power; they only store energy, so placing a capacitor in parallel with our load will not change the power. We also see that the power factor is 1, so we accomplished our goal of creating a new load with a unity power factor.

EXERCISE 5-10: Calculate the power absorbed by the load and the power factor of the load in the circuit below. Assume that the frequency is 50 Hz and the voltage is 230 VRMS.

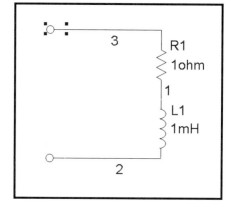

SOLUTION: Add an AC voltage source to the circuit and set the voltage to 230 VRMS and the frequency to 50 Hz. Add the Wattmeter and wire it as shown. The results are displayed on the meter window:

For power factor correction, we need to add a 910.2 µF capacitor in parallel with the load:

5.G. Measuring Impedance

The technique presented here can be used to measure the impedance or resistance between any two nodes. For purely resistive circuits, we could use the Multimeter. However, the Multimeter cannot measure the impedance of a circuit with capacitances or inductances, nor can it measure the input or output impedance of a circuit that contains active components such as transistors or OPAMPs. We will find the AC impedance between two nodes. We will illustrate using two examples. The first will be a passive circuit with resistors only. The second will be a jFET source follower. For the passive circuit, we will first measure the resistance using the Multimeter so that we can compare the result to our new measurement technique.

5.G.1. Resistive Measurement with the Multimeter

We will find the impedance between nodes **1** and **0** in the circuit below:

Since this is a resistive circuit, we can add the Multimeter instrument and measure the impedance. Click on the Multimeter button ![btn] as shown to place the instrument, and wire it to your circuit as shown:

Double-click the **LEFT** mouse button on the instrument to open its window:

The last thing we must do is ground our circuit. Although it would not be necessary in the lab to ground a circuit before using the Multimeter to measure a resistance, it is necessary with MultiSim because Multisim must have a ground reference in order to calculate voltages in a circuit. Add a ground symbol as shown:

We can now run the simulation. Click the **LEFT** mouse button on the Run Simulation button ![btn] to start the simulation:

EWB Multisim   Magnitude and Phase Simulations   271

By default, the Multimeter displays the voltage between the two test leads. We need to change this to resistance, so click the **LEFT** mouse button on the Ω button to measure the resistance:

After a few moments the meter indicates that the resistance is 2 kΩ, which is the expected result. Click on the Run Simulation button again to stop the simulation.

Using the Multimeter is a very easy way to measure resistance, but it can only measure the resistance of a network of resistors. To measure anything more complicated, we will need to use the technique presented in the next section.

5.G.2. Impedance Measurement of a Passive Circuit Using SPICE

We will now use a new measurement technique to find the resistance of the resistor network used in the previous section. We will use the same circuit here so that we can show that the two methods yield the same answer. We will once again measure the resistance of the circuit below:

Add an AC current source as shown below. The test current source is added between nodes 1 and 0 because we are interested in the impedance between those two nodes. Either an AC current or an AC voltage source can be used. Wire the circuit shown. Note that ground now serves as node 0.

Although this circuit has no dependent sources, inductors, or capacitors, the circuit could contain any circuit element and we could use this method to find the impedance. This circuit was chosen because the impedance is easily calculated and can be compared to this result or the result obtained with the Multimeter instrument. For this circuit, the calculated impedance between nodes **1** and **0** is 2 kΩ.

Next we need to set up and run the AC Analysis. Select **Simulate**, **Analyses**, and then **AC Analysis** from the menus:

Since this is a resistive circuit, the frequency settings are not important and we will leave them set to their default settings. The one parameter we will change is the **Vertical scale**. Since this is a passive circuit and the resistance will not change with frequency, we will change the vertical scale to **Linear**:

The resistance of our circuit is the voltage between node 1 and ground divided by the magnitude of the test current source, V(1)/I(I_test). Since we did not specify a value for the AC current source, its value will be 1 A. With this value of current, the resistance of the network is numerically equal to the voltage at node 1. Thus, we only need to plot the voltage at node 1.

The last thing we need to do is specify the output variables. We are only interested in the voltage at Node 1, so select the **Output variables** tab and then specify to plot node 1 during the simulation:

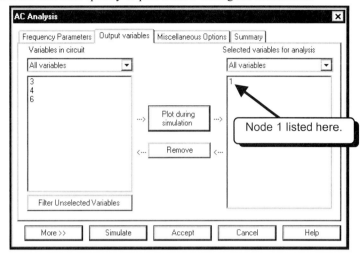

Click on the **Simulate** button to run the simulation. You will see this trace:

Use the cursors to find a numerical value:

This trace shows that the impedance is constant with frequency and the y-coordinates of the cursors show that the resistance is at the expected value of 2 kΩ. If we had inductors or capacitors in the circuit, the impedance would have changed with frequency.

EXERCISE 5-11: Find the equivalent impedance of the following resistor network:

SOLUTION: Add an independent current source between the two terminals where you wish to measure the impedance:

The equivalent impedance is V(**1**)/I(**Itest**). Since **Itest** is a 1-amp source, we can just plot the voltage at node 1. Set up and run an AC Analysis and plot the voltage at node 1 during the simulation:

The impedance is **2.2394** kΩ.

5.G.3. Impedance Measurement of an Active Circuit Using SPICE

We would now like to find the output impedance of the jFET source follower circuit shown below. The output is taken at the source terminal of the jFET, so the impedance we are interested in is between nodes **vo** and **0** (ground). Although this is a jFET circuit, we can use this technique with any circuit.

When you are finding output impedance you must set the input AC source to zero. Since the input is a voltage source, we replace **V1** with a short:

Next we need to add an AC source at the output to measure the impedance. We could use either an AC voltage source or an AC current source. If you use a voltage source, wire the circuit shown below:

Capacitor **C2** is necessary to preserve the bias of the amplifier. Remember that **V1** is an AC source. For biasing, all AC sources are set to zero. Since **V1** is a voltage source, it would be replaced by a short. Without **C2**, the source terminal of the jFET would be grounded when calculating the bias. This would destroy the bias and render the impedance measurement invalid.

The output impedance of the source follower can be observed by plotting the ratio V(Vo)/I(V1) in the circuit above. It is important to note that the ratio V(Vx)/I(V1) is not the output impedance of the source follower, but the output impedance plus the impedance of the capacitor.

The second method to measure the impedance of a circuit uses an AC current source:

I1 is an AC current source. Note that there is no blocking capacitor between the current source and the amplifier. When the bias is calculated, all AC sources are set to zero. When a current source is set to zero, it is replaced by an open circuit. An open circuit at **Vo** is equivalent to the original circuit for bias calculations, and the bias is preserved. The output impedance of the circuit can be calculated by the ratio V(Vo)/I(I1). Also note that since the magnitude of I1 is 1 amp, the impedance is numerically equal to the voltage at node **Vo**.

We next need to set up an AC Analysis and then plot the voltage at node Vo. Select **Simulate**, **Analyses**, and then **AC Analysis** from the menus, and then fill in the dialog box as shown:

Select the **Output variables** tab and specify to plot **Vo** during the simulation:

EWB Multisim — Magnitude and Phase Simulations

We are now ready to run the simulation. Click on the **Simulate** button and then use the cursors to obtain a numerical value of the output resistance:

The trace shows that the impedance is about 42 Ω for frequencies up to 1 MHz.

EXERCISE 5-12: Find the output impedance of the emitter-follower:

SOLUTION: Replace the independent voltage source with a short circuit and add an independent current source between the two terminals where you wish to measure the impedance:

The output impedance is V(Vo)/I(I1). Since I1 has a magnitude of 1, the output impedance is numerically equal to the voltage at **Vo**:

At mid-band, the impedance is about 2.2 Ω. The change in impedance at low frequencies is due to **Cs**. The change in impedance at high frequencies is due to capacitances internal to the transistor.

5.H. Problems

Problem 5-1: Find the RMS magnitude of the voltage at node V1 using indicators in the circuit below:

Problem 5-2: For the circuit of Problem 5-1, find the RMS magnitude of the voltage at node V2 and the current through R1 using indicators.

Problem 5-3: For the circuit of Problem 5-1, find the RMS magnitude of the voltage at node V3 and the current through C2 using indicators.

Problem 5-4: For the circuit of Problem 5-1, find the RMS magnitude of the voltage across R1 and the current through C2 using indicators.

Problem 5-5: For the circuit of Problem 5-1, find the amplitude (peak value) of the voltage at node V1 and the current through R2 using indicators.

Problem 5-6: For the circuit of Problem 5-1, find the amplitude (peak value) of the voltage at node V2 and the current through C1 using indicators.

Problem 5-7: For the circuit of Problem 5-1, find the amplitude (peak value) of the voltage at node V3 and the current through R3 using indicators.

Problem 5-8: For the circuit of Problem 5-1, find the RMS magnitude of the voltage across C2 and the current through C2 using indicators.

Problem 5-9: For the circuit of Problem 5-1, find the RMS magnitude of the voltage at node V1 and the current through R1 using Multimeters.

Problem 5-10: For the circuit of Problem 5-1, find the RMS magnitude of the voltage at node V2 and the current through R2 using Multimeters.

Problem 5-11: For the circuit of Problem 5-1, find the RMS magnitude of the voltage at node V3 and the current through R3 using Multimeters.

Problem 5-12: For the circuit of Problem 5-1, find the RMS magnitude of the voltage across C1 and the current through R2 using Multimeters.

Problem 5-13: For the circuit of Problem 5-1, find the amplitude (peak value) of the voltage at node V1 and the current through C2 using Multimeters.

Problem 5-14: For the circuit of Problem 5-1, find the amplitude (peak value) of the voltage at node V2 and the current through C1 using Multimeters.

Problem 5-15: For the circuit of Problem 5-1, find the amplitude (peak value) of the voltage at node V3 and the current through C2 using Multimeters.

Problem 5-16: For the circuit of Problem 5-1, find the amplitude (peak value) of the voltages at nodes V1, V2, and V3 using the SPICE AC Sweep analysis.

Problem 5-17: For the circuit of Problem 5-1, find the phase of the voltages at nodes V1, V2, and V3 using the SPICE AC Sweep analysis.

Problem 5-18: Find the RMS magnitude of the voltage at node V1 using indicators in the circuit below:

Problem 5-19: For the circuit of Problem 5-18, find the RMS magnitude of the voltage at node V2 and the current through R1 using indicators.

Problem 5-20: For the circuit of Problem 5-18, find the RMS magnitude of the voltage at node V3 and the current through C2 using indicators.

Problem 5-21: For the circuit of Problem 5-18, find the amplitude (peak value) of the voltage at node V1 and the current through R2 using indicators.

Problem 5-22: For the circuit of Problem 5-18, find the amplitude (peak value) of the voltage at node V2 and the current through C1 using indicators.

Problem 5-23: For the circuit of Problem 5-18, find the amplitude (peak value) of the voltage across C2 and the current through C1 using indicators.

Problem 5-24: For the circuit of Problem 5-18, find the amplitude (peak value) of the voltage at node V3 and the current through L1 using indicators.

Problem 5-25: For the circuit of Problem 5-18, find the RMS magnitude of the voltage at node V1 and the current through R1 using Multimeters.

Problem 5-26: For the circuit of Problem 5-18, find the RMS magnitude of the voltage at node V2 and the current through R2 using Multimeters.

Problem 5-27: For the circuit of Problem 5-18, find the RMS magnitude of the voltage at node V3 and the current through L1 using Multimeters.

Problem 5-28: For the circuit of Problem 5-18, find the RMS magnitude of the voltage across R3 and the current through L1 using Multimeters.

Problem 5-29: For the circuit of Problem 5-18, find the amplitude (peak value) of the voltage at node V1 and the current through C2 using Multimeters.

Problem 5-30: For the circuit of Problem 5-18, find the amplitude (peak value) of the voltage at node V2 and the current through C1 using Multimeters.

Problem 5-31: For the circuit of Problem 5-18, find the amplitude (peak value) of the voltage at node V3 and the current through L1 using Multimeters.

Problem 5-32: For the circuit of Problem 5-18, find the amplitude (peak value) of the voltage across C1 and the current through L1 using Multimeters.

Problem 5-33: For the circuit of Problem 5-18, find the amplitude (peak value) of the voltages at nodes V1, V2, and V3 using the SPICE AC Sweep analysis.

Problem 5-34: For the circuit of Problem 5-18, find the phase of the voltages at nodes V1, V2, and V3 using the SPICE AC Sweep analysis.

Problem 5-35: Find the RMS magnitude of the voltage at node V1 using indicators in the circuit below:

Problem 5-36: For the circuit of Problem 5-35, find the RMS magnitude of the voltage at node V2 and the current through R1 using indicators.

Problem 5-37: For the circuit of Problem 5-35, find the RMS magnitude of the voltage at node V3 and the current through C2 using indicators.

Problem 5-38: For the circuit of Problem 5-35, find the amplitude (peak value) of the voltage at node V1 and the current through R2 using indicators.

Problem 5-39: For the circuit of Problem 5-35, find the amplitude (peak value) of the voltage across L1 and the current through R2 using indicators.

Problem 5-40: For the circuit of Problem 5-35, find the amplitude (peak value) of the voltage at node V2 and the current through C1 using indicators.

Problem 5-41: For the circuit of Problem 5-35, find the amplitude (peak value) of the voltage at node V3 and the current through L1 using indicators.

Problem 5-42: For the circuit of Problem 5-35, find the RMS magnitude of the voltage at node V1 and the current through R1 using Multimeters.

Problem 5-43: For the circuit of Problem 5-35, find the RMS magnitude of the voltage at node V2 and the current through R2 using Multimeters.

Problem 5-44: For the circuit of Problem 5-35, find the RMS magnitude of the voltage across C2 and the current through R2 using Multimeters.

Problem 5-45: For the circuit of Problem 5-35, find the RMS magnitude of the voltage at node V3 and the current through L1 using Multimeters.

Problem 5-46: For the circuit of Problem 5-35, find the amplitude (peak value) of the voltage at node V1 and the current through C2 using Multimeters.

Problem 5-47: For the circuit of Problem 5-35, find the amplitude (peak value) of the voltage at node V2 and the current through C1 using Multimeters.

Problem 5-48: For the circuit of Problem 5-35, find the amplitude (peak value) of the voltage at node V3 and the current through L1 using Multimeters.

Problem 5-49: For the circuit of Problem 5-35, find the amplitude (peak value) of the voltages at nodes V1, V2, and V3 using the SPICE AC Sweep analysis.

Problem 5-50: For the circuit of Problem 5-35, find the phase of the voltages at nodes V1, V2, and V3 using the SPICE AC Sweep analysis.

Problem 5-51: For the circuit below, generate the Bode magnitude and phase plots of Vo/Vin using the Bode Plotter instrument. Use the cursors to find the frequency where the gain is down by –3 dB.

Problem 5-52: For the circuit of Problem 5-51, generate the Bode magnitude and phase plots of Vo/Vin using the SPICE AC Sweep analysis. Use the cursors to find the frequency where the gain is down by –3 dB. What are the gain and phase at this point?

Problem 5-53: For the circuit of Problem 5-51, generate the Bode magnitude and phase plots of Vo/Vin using the SPICE AC Sweep analysis. Use the cursors to find the frequency where the gain is down by –20 dB. What are the gain and phase at this point?

Problem 5-54: For the circuit of Problem 5-51, generate the Bode magnitude and phase plots of Vo/Vin using the SPICE AC Sweep analysis. Use the cursors to find the frequency where the gain is down by –40 dB. What are the gain and phase at this point?

Problem 5-55: For the circuit of Problem 5-51, generate the Bode magnitude and phase plots of Vo/Vin using the SPICE AC Sweep analysis. Use the cursors to find the frequency where the gain is down by –60 dB. What are the gain and phase at this point?

Problem 5-56: For the circuit of Problem 5-51, generate the Bode magnitude and phase plots of Vo/Vin using the SPICE AC Sweep analysis. Use the cursors to find the frequency where the phase is –45°. What are the gain and phase at this point?

Problem 5-57: For the circuit of Problem 5-51, generate the Bode magnitude and phase plots of Vo/Vin using the SPICE AC Sweep analysis. Use the cursors to find the frequency where the phase is –22°. What are the gain and phase at this point?

Problem 5-58: For the circuit of Problem 5-51, generate the Bode magnitude and phase plots of Vo/Vin using the SPICE AC Sweep analysis. Use the cursors to find the frequency where the phase is –75°. What are the gain and phase at this point?

Problem 5-59: For the circuit of Problem 5-51, generate the Bode magnitude and phase plots of Vo/Vin using the Bode Plotter instrument. Use the cursors to find the frequency where the gain is down by –20 dB. What are the gain and phase at this point?

Problem 5-60: For the circuit of Problem 5-51, generate the Bode magnitude and phase plots of Vo/Vin using the Bode Plotter instrument. Use the cursors to find the frequency where the gain is down by –40 dB. What are the gain and phase at this point?

Problem 5-61: For the circuit of Problem 5-51, generate the Bode magnitude and phase plots of Vo/Vin using the Bode Plotter instrument. Use the cursors to find the frequency where the gain is down by –60 dB. What are the gain and phase at this point?

Problem 5-62: For the circuit of Problem 5-51, generate the Bode magnitude and phase plots of Vo/Vin using the Bode Plotter instrument. Use the cursors to find the frequency where the phase is –45°. What are the gain and phase at this point?

Problem 5-63: For the circuit of Problem 5-51, generate the Bode magnitude and phase plots of Vo/Vin using the Bode Plotter instrument. Use the cursors to find the frequency where the phase is –22°. What are the gain and phase at this point?

Problem 5-64: For the circuit of Problem 5-51, generate the Bode magnitude and phase plots of Vo/Vin using the Bode Plotter instrument. Use the cursors to find the frequency where the phase is –75°. What are the gain and phase at this point?

Problem 5-65: For the circuit below, generate the Bode magnitude and phase plots of Vo/Vin using the Bode Plotter instrument. Use the cursors to find the frequencies where the gain is down by –3 dB from the midband gain. Simulate the circuit for frequencies from 10 mHz to 10 GHz.

Problem 5-66: For the circuit of Problem 5-65, generate the Bode magnitude and phase plots of Vo/Vin using the SPICE AC Sweep analysis. Use the cursors to find the frequencies where the gain is down by –3 dB from the midband gain. What are the gain and phase at this point? Simulate the circuit for frequencies from 10 mHz to 10 GHz.

Problem 5-67: For the circuit of Problem 5-65, generate the Bode magnitude and phase plots of Vo/Vin using the SPICE AC Sweep analysis. Use the cursors to find the frequencies where the gain is down by –20 dB from the midband gain. What are the gain and phase at this point? Simulate the circuit for frequencies from 10 mHz to 10 GHz.

Problem 5-68: For the circuit of Problem 5-65, generate the Bode magnitude and phase plots of Vo/Vin using the SPICE AC Sweep analysis. Use the cursors to find the frequencies where the gain is down by –40 dB from the midband gain. What are the gain and phase at this point? Simulate the circuit for frequencies from 10 mHz to 10 GHz.

Problem 5-69: For the circuit of Problem 5-65, generate the Bode magnitude and phase plots of Vo/Vin using the SPICE AC Sweep analysis. Use the cursors to find the frequencies where the gain is down by –60 dB from the midband gain. What are the gain and phase at this point? Simulate the circuit for frequencies from 10 mHz to 10 GHz.

Problem 5-70: For the circuit of Problem 5-65, generate the Bode magnitude and phase plots of Vo/Vin using the SPICE AC Sweep analysis. Use the cursors to find the frequency where the phase is –45°. What are the gain and phase at this point? Simulate the circuit for frequencies from 10 mHz to 10 GHz.

Problem 5-71: For the circuit of Problem 5-65, generate the Bode magnitude and phase plots of Vo/Vin using the SPICE AC Sweep analysis. Use the cursors to find the frequency where the phase is +22°. What are the gain and phase at this point? Simulate the circuit for frequencies from 10 mHz to 10 GHz.

Problem 5-72: For the circuit of Problem 5-65, generate the Bode magnitude and phase plots of Vo/Vin using the SPICE AC Sweep analysis. Use the cursors to find the frequency where the phase is +75°. What are the gain and phase at this point? Simulate the circuit for frequencies from 10 mHz to 10 GHz.

Problem 5-73: For the circuit of Problem 5-65, generate the Bode magnitude and phase plots of Vo/Vin using the Bode Plotter instrument. Use the cursors to find the frequencies where the gain is down by –20 dB from the midband gain. What are the gain and phase at this point? Simulate the circuit for frequencies from 10 mHz to 10 GHz.

Problem 5-74: For the circuit of Problem 5-65, generate the Bode magnitude and phase plots of Vo/Vin using the Bode Plotter instrument. Use the cursors to find the frequencies where the gain is down by –40 dB from the midband gain. What are the gain and phase at this point? Simulate the circuit for frequencies from 10 mHz to 10 GHz.

Problem 5-75: For the circuit of Problem 5-65, generate the Bode magnitude and phase plots of Vo/Vin using the Bode Plotter instrument. Use the cursors to find the frequencies where the gain is down by –60 dB from the midband gain. What are the gain and phase at this point? Simulate the circuit for frequencies from 10 mHz to 10 GHz.

Problem 5-76: For the circuit of Problem 5-65, generate the Bode magnitude and phase plots of Vo/Vin using the Bode Plotter instrument. Use the cursors to find the frequency where the phase is +45°. What are the gain and phase at this point? Simulate the circuit for frequencies from 10 mHz to 10 GHz.

Problem 5-77: For the circuit of Problem 5-65, generate the Bode magnitude and phase plots of Vo/Vin using the Bode Plotter instrument. Use the cursors to find the frequency where the phase is –22°. What are the gain and phase at this point? Simulate the circuit for frequencies from 10 mHz to 10 GHz.

Problem 5-78: For the circuit of Problem 5-65, generate the Bode magnitude and phase plots of Vo/Vin using the Bode Plotter instrument. Use the cursors to find the frequency where the phase is −75°. What are the gain and phase at this point? Simulate the circuit for frequencies from 10 mHz to 10 GHz.

Problem 5-79: For the circuit below, use the Bode Plotter to generate a magnitude plot and find the midband gain (Vout/Vin), and the upper and lower −3 dB frequencies for this amplifier. (Note that this is nearly the same circuit used in section 5.C. The values of the components are the same.)

Problem 5-80: Repeat Problem 5-79 using the SPICE AC Sweep analysis.

Problem 5-81: Repeat the simulations of Problem 5-79 with RL = 10 kΩ, RL = 100 Ω, and RL = 10 Ω.

Problem 5-82: Repeat the simulations of Problem 5-79 with RC = 0 Ω. What is the effect of RC on this circuit?

Problem 5-83: For the circuit below, use the Bode Plotter to generate a magnitude plot and find the midband gain (Vout/Vin), and the upper and lower −3 dB frequencies for this amplifier. (Note that this is nearly the same circuit used in section 5.C. The values of the components are the same.)

Problem 5-84: Repeat Problem 5-83 using the SPICE AC Sweep analysis.

Problem 5-85: Repeat the simulations of Problem 5-83 with RL = 10 kΩ, RL = 100 Ω, and RL = 10 Ω.

Problem 5-86: For the circuit below, use the Bode Plotter to generate a magnitude plot and find the midband gain (Vout/Vin), and the upper and lower −3 dB frequencies for this amplifier. (Note that this is nearly the same circuit used in section 5.C. The values of the components are the same.)

Problem 5-87: Repeat Problem 5-86 using the SPICE AC Sweep analysis.

Problem 5-88: Repeat the simulations of Problem 5-86 with RL = 1 kΩ, RL = 100 Ω, and RL = 10 Ω.

Problem 5-89: For the circuit below, use the Bode Plotter to generate a magnitude plot and find the midband gain (Vout/Vin), and the upper and lower −3 dB frequencies for this amplifier. (Note that this is nearly the same circuit used in section 5.C. The values of the components are the same, and CE has been removed.)

Problem 5-90: Repeat Problem 5-89 using the SPICE AC Sweep analysis.

Problem 5-91: Repeat the simulations of Problem 5-89 with RL = 10 kΩ, RL = 100 Ω, and RL = 10 Ω.

Problem 5-92:

(a) Design a non-inverting OPAMP circuit with a gain of 1 using a LM741CH. Create a Bode magnitude plot, verify the midband gain, and find the upper −3 dB frequency.

(b) Design an inverting OPAMP circuit with a gain of −1 using an LM741CH. Create a Bode magnitude plot, verify the midband gain, and find the upper −3 dB frequency. Use a minimum resistor size of 10 kΩ.

(c) Although the circuits from parts (a) and (b) have the same magnitude of gain, the upper −3 dB frequency is different. Why?

Problem 5-93:

(a) Design a non-inverting OPAMP circuit with a gain of 1 using a TL072ACD. Create a Bode magnitude plot, verify the midband gain, and find the upper −3 dB frequency.

(b) Design an inverting OPAMP circuit with a gain of −1 using a TL072ACD. Create a Bode magnitude plot, verify the midband gain, and find the upper −3 dB frequency. Use a minimum resistor size of 10 kΩ.

(c) Although the circuits from parts (a) and (b) have the same magnitude of gain, the upper −3 dB frequency is different. Why?

Problem 5-94:

(a) Design a non-inverting OPAMP circuit with a gain of 1 using an OPA602BM. Create a Bode magnitude plot, verify the midband gain, and find the upper −3 dB frequency.

(b) Design an inverting OPAMP circuit with a gain of −1 using an OPA602BM. Create a Bode magnitude plot, verify the midband gain, and find the upper −3 dB frequency. Use a minimum resistor size of 10 kΩ.

(c) Although the circuits from parts (a) and (b) have the same magnitude of gain, the upper −3 dB frequency is different. Why?

Problem 5-95:

(a) Design a non-inverting OPAMP circuit with a gain of 100 using an LM741CH. Create a Bode magnitude plot, verify the midband gain, and find the upper −3 dB frequency.

(b) Design an inverting OPAMP circuit with a gain of −100 using an LM741CH. Create a Bode magnitude plot, verify the midband gain, and find the upper −3 dB frequency.

(c) The simulations from parts (a) and (b) show the same gain and upper −3 dB frequency. This is different than in Problem 5-92. Why?

Problem 5-96:

(a) Design a non-inverting OPAMP circuit with a gain of 100 using a TL072ACD. Create a Bode magnitude plot, verify the midband gain, and find the upper −3 dB frequency.

(b) Design an inverting OPAMP circuit with a gain of −100 using a TL072ACD. Create a Bode magnitude plot, verify the midband gain, and find the upper −3 dB frequency.

(c) The simulations from parts (a) and (b) show the same gain and upper −3 dB frequency. This is different than in Problem 5-93. Why?

Problem 5-97:

(a) Design a non-inverting OPAMP circuit with a gain of 100 using an OPA602BM. Create a Bode magnitude plot, verify the midband gain, and find the upper −3 dB frequency.

(b) Design an inverting OPAMP circuit with a gain of −100 using an OPA602BM. Create a Bode magnitude plot, verify the midband gain, and find the upper −3 dB frequency.

(c) The simulations from parts (a) and (b) show the same gain and upper −3 dB frequency. This is different than in Problem 5-94. Why?

Problem 5-98: Design an inverting summing amplifier that implements the equation $V_{OUT} = -V_A - 10V_B - 100V_C$ using your favorite OPAMP. (Do not use the virtual OPAMPS.)

(a) With the inputs to V_B and V_C set to zero (grounded) obtain a Bode plot of V_{OUT}/V_A. Verify the midband gain and find the upper −3 dB frequency.

(b) With the inputs to V_A and V_C set to zero (grounded) obtain a Bode plot of V_{OUT}/V_B. Verify the midband gain and find the upper −3 dB frequency.

(c) With the inputs to V_A and V_B set to zero (grounded) obtain a Bode plot of V_{OUT}/V_C. Verify the midband gain and find the upper −3 dB frequency.

(d) Use the Postprocessor to generate Bode magnitude plots of V_{OUT}/V_A, V_{OUT}/V_B, and V_{OUT}/V_C on a single graph.

(e) Although the gain from each input is different, the upper −3 dB frequencies for each plot generated above are the same. Why?

Problem 5-99: For the circuit below, use the Parameter Sweep to plot the gain V_{OUT}/I_{IN} in decibels for values of RF equal to 1 kΩ, 10 kΩ, 100 kΩ, and 1 MΩ.

Problem 5-100: For the circuit below, use the Parameter Sweep to plot the gain V_{OUT}/V_{IN} in decibels for values of R4 equal to 1 kΩ, 10 kΩ, 100 kΩ, and 1 MΩ.

Problem 5-101: A complex load has an impedance of $15\angle -37°$ at 60 Hz.

(a) Create a circuit of inductors, capacitors, and/or resistors that implements this impedance. You can use as many or as few components as you like.

(b) By hand, calculate the power factor. Does this circuit have a leading or lagging power factor?

(c) Calculate by hand the amount of power this load will absorb in watts if it is connected to a 115 V_{RMS} 60 Hz source.

(d) Verify with Multisim the power factor and amount of power this load absorbs.

(e) Specify a single component that can be added to the load to make the compensated load have a unity power factor.

(f) Calculate by hand the amount of power the compensated load will absorb in watts if it is connected to a 115 V_{RMS} 60 Hz source.

(g) Verify with Multisim the power factor and amount of power the compensated load absorbs.

Problem 5-102: A complex load has an impedance of $15\angle 37°$ at 60 Hz.

(a) Create a circuit of inductors, capacitors, and/or resistors that implements this impedance. You can use as many or as few components as you like.

(b) By hand, calculate the power factor. Does this circuit have a leading or lagging power factor?

(c) Calculate by hand the amount of power this load will absorb in watts if it is connected to a 115 V_{RMS} 60 Hz source.

(d) Verify with Multisim the power factor and amount of power this load absorbs.

(e) Specify a single component that can be added to the load to make the compensated load have a unity power factor.

(f) Calculate by hand the amount of power the compensated load will absorb in watts if it is connected to a 115 V_{RMS} 60 Hz source.

(g) Verify with Multisim the power factor and amount of power the compensated load absorbs.

Problem 5-103: A complex load has an impedance of $25\angle -25°$ at 400 Hz.

(a) Create a circuit of inductors, capacitors, and/or resistors that implements this impedance. You can use as many or as few components as you like.

(b) By hand, calculate the power factor. Does this circuit have a leading or lagging power factor?

(c) Calculate by hand the amount of power this load will absorb in watts if it is connected to a 220 V_{RMS} 400 Hz source.

(d) Verify with Multisim the power factor and amount of power this load absorbs.

(e) Specify a single component that can be added to the load to make the compensated load have a unity power factor.

(f) Calculate by hand the amount of power the compensated load will absorb in watts if it is connected to a 220 V_{RMS} 400 Hz source.

(g) Verify with Multisim the power factor and amount of power the compensated load absorbs.

Problem 5-104: A complex load has an impedance of $25\angle 25°$ at 400 Hz.

(a) Create a circuit of inductors, capacitors, and/or resistors that implements this impedance. You can use as many or as few components as you like.

(b) By hand, calculate the power factor. Does this circuit have a leading or lagging power factor?

(c) Calculate by hand the amount of power this load will absorb in watts if it is connected to a 220 V_{RMS} 400 Hz source.

(d) Verify with Multisim the power factor and amount of power this load absorbs.

(e) Specify a single component that can be added to the load to make the compensated load have a unity power factor.

(f) Calculate by hand the amount of power the compensated load will absorb in watts if it is connected to a 220 V_{RMS} 400 Hz source.

(g) Verify with Multisim the power factor and amount of power the compensated load absorbs.

Problem 5-105: A complex load has an impedance of $1\angle -30°$ at 60 Hz.

(a) Create a circuit of inductors, capacitors, and/or resistors that implements this impedance. You can use as many or as few components as you like.

(b) By hand, calculate the power factor. Does this circuit have a leading or lagging power factor?

(c) Calculate by hand the amount of power this load will absorb in watts if it is connected to a 115 V_{RMS} 60 Hz source.

(d) Verify with Multisim the power factor and amount of power this load absorbs.

(e) Specify a single component that can be added to the load to make the compensated load have a unity power factor.

(f) Calculate by hand the amount of power the compensated load will absorb in watts if it is connected to a 115 V_{RMS} 60 Hz source.

(g) Verify with Multisim the power factor and amount of power the compensated load absorbs.

Problem 5-106: A complex load has an impedance of $1\angle 30°$ at 60 Hz.

(a) Create a circuit of inductors, capacitors, and/or resistors that implements this impedance. You can use as many or as few components as you like.

(b) By hand, calculate the power factor. Does this circuit have a leading or lagging power factor?

(c) Calculate by hand the amount of power this load will absorb in watts if it is connected to a 115 V_{RMS} 60 Hz source.

(d) Verify with Multisim the power factor and amount of power this load absorbs.

(e) Specify a single component that can be added to the load to make the compensated load have a unity power factor.

(f) Calculate by hand the amount of power the compensated load will absorb in watts if it is connected to a 115 V_{RMS} 60 Hz source.

(g) Verify with Multisim the power factor and amount of power the compensated load absorbs.

Problem 5-107: A complex load has an impedance of $0.5\angle -20°$ at 400 Hz.

(a) Create a circuit of inductors, capacitors, and/or resistors that implements this impedance. You can use as many or as few components as you like.

(b) By hand, calculate the power factor. Does this circuit have a leading or lagging power factor?

(c) Calculate by hand the amount of power this load will absorb in watts if it is connected to a 220 V_{RMS} 400 Hz source.

(d) Verify with Multisim the power factor and amount of power this load absorbs.

(e) Specify a single component that can be added to the load to make the compensated load have a unity power factor.

(f) Calculate by hand the amount of power the compensated load will absorb in watts if it is connected to a 220 V_{RMS} 400 Hz source.

(g) Verify with Multisim the power factor and amount of power the compensated load absorbs.

Problem 5-108: A complex load has an impedance of $0.5\angle 20°$ at 400 Hz.

(a) Create a circuit of inductors, capacitors, and/or resistors that implements this impedance. You can use as many or as few components as you like.

(b) By hand, calculate the power factor. Does this circuit have a leading or lagging power factor?

(c) Calculate by hand the amount of power this load will absorb in watts if it is connected to a 220 V_{RMS} 400 Hz source.

(d) Verify with Multisim the power factor and amount of power this load absorbs.

(e) Specify a single component that can be added to the load to make the compensated load have a unity power factor.

(f) Calculate by hand the amount of power the compensated load will absorb in watts if it is connected to a 220 V_{RMS} 400 Hz source.

(g) Verify with Multisim the power factor and amount of power the compensated load absorbs.

Problem 5-109: A load that is connected to a 120 V_{RMS} 60 Hz source has a power factor of 0.93 lagging and absorbs 360 watts of power.

(a) What is the impedance of the load?

(b) Create a circuit of inductors, capacitors, and/or resistors that implements this impedance. You can use as many or as few components as you like.

(c) By hand, calculate the power factor. Does this circuit have a leading or lagging power factor?

(d) Calculate by hand the amount of power this load will absorb in watts if it is connected to a 220 V_{RMS} 400 Hz source.

(e) Verify with Multisim the power factor and amount of power this load absorbs.

(f) Specify a single component that can be added to the load to make the compensated load have a unity power factor.

(g) Calculate by hand the amount of power the compensated load will absorb in watts if it is connected to a 220 V_{RMS} 400 Hz source.

(h) Verify with Multisim the power factor and amount of power the compensated load absorbs.

Problem 5-110: A load that is connected to a 220 V_{RMS} 60 Hz source has a power factor of 0.93 leading and absorbs 1200 Watts of power.

(a) What is the impedance of the load?

(b) Create a circuit of inductors, capacitors, and/or resistors that implements this impedance. You can use as many or as few components as you like.

(c) By hand, calculate the power factor. Does this circuit have a leading or lagging power factor?

(d) Calculate by hand the amount of power this load will absorb in watts if it is connected to a 220 V_{RMS} 400 Hz source.

(e) Verify with Multisim the power factor and amount of power this load absorbs.

(f) Specify a single component that can be added to the load to make the compensated load have a unity power factor.

(g) Calculate by hand the amount of power the compensated load will absorb in watts if it is connected to a 220 V_{RMS} 400 Hz source.

(h) Verify with Multisim the power factor and amount of power the compensated load absorbs.

Problem 5-111: A load that is connected to a 208 V_{RMS} 400 Hz source has a power factor of 0.89 leading and absorbs 10 kW of power.

(a) What is the impedance of the load?

(b) Create a circuit of inductors, capacitors, and/or resistors that implements this impedance. You can use as many or as few components as you like.

(c) By hand, calculate the power factor. Does this circuit have a leading or lagging power factor?

(d) Calculate by hand the amount of power this load will absorb in watts if it is connected to a 220 V_{RMS} 400 Hz source.

(e) Verify with Multisim the power factor and amount of power this load absorbs.

(f) Specify a single component that can be added to the load to make the compensated load have a unity power factor.

(g) Calculate by hand the amount of power the compensated load will absorb in watts if it is connected to a 220 V_{RMS} 400 Hz source.

(h) Verify with Multisim the power factor and amount of power the compensated load absorbs.

Problem 5-112: A load that is connected to a 208 V_{RMS} 400 Hz source has a power factor of 0.85 lagging and absorbs 5 kW of power.

(a) What is the impedance of the load?

(b) Create a circuit of inductors, capacitors, and/or resistors that implements this impedance. You can use as many or as few components as you like.

(c) By hand, calculate the power factor. Does this circuit have a leading or lagging power factor?

(d) Calculate by hand the amount of power this load will absorb in watts if it is connected to a 220 V_{RMS} 400 Hz source.

(e) Verify with Multisim the power factor and amount of power this load absorbs.

(f) Specify a single component that can be added to the load to make the compensated load have a unity power factor.

(g) Calculate by hand the amount of power the compensated load will absorb in watts if it is connected to a 220 V_{RMS} 400 Hz source.

(h) Verify with Multisim the power factor and amount of power the compensated load absorbs.

Problem 5-113: Find the impedance of the circuit below using the Multimeter.

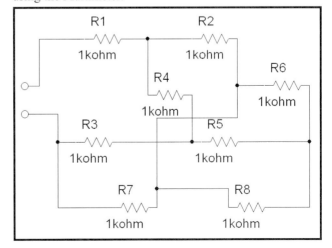

Problem 5-114: Find the impedance of the circuit below using the Multimeter.

Problem 5-115: Find the impedance of the circuit below using the Multimeter.

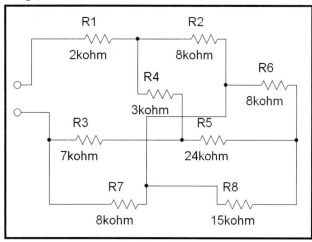

Problem 5-116: Find the impedance of the circuit below using the Multimeter.

Problem 5-117: Find the impedance of the circuit below using the Multimeter.

Problem 5-118: Find the impedance of the circuit below using the Multimeter.

Problem 5-119: Find the impedance of the circuit below using the Multimeter.

Problem 5-120: Find the impedance of the circuit below using the Multimeter.

Problem 5-121: Find the impedance of the circuit below using the Multimeter.

Problem 5-122: Find the impedance of the circuit below using the AC Sweep analysis.

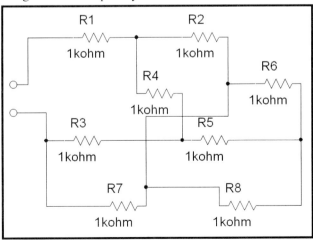

Problem 5-123: Find the impedance of the circuit below using the AC Sweep analysis.

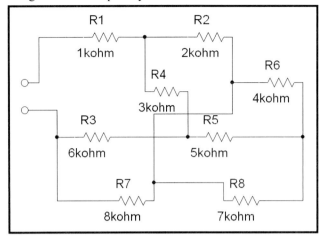

Problem 5-124: Find the impedance of the circuit below using the AC Sweep analysis.

Problem 5-125: Find the impedance of the circuit below using the AC Sweep analysis.

Problem 5-126: Find the impedance of the circuit below using the AC Sweep analysis.

Problem 5-127: Find the impedance of the circuit below using the AC Sweep analysis.

Problem 5-128: Find the impedance of the circuit below using the AC Sweep analysis.

Problem 5-129: Find the impedance of the circuit below using the AC Sweep analysis.

Problem 5-130: Find the impedance of the circuit below using the AC Sweep analysis.

Problem 5-131: For the circuit below, find the AC output impedance between node V_{OUT} and ground using the SPICE AC Analysis.

Problem 5-132: For the circuit below, find the AC input impedance between node V_{IN} and ground using the SPICE AC Analysis.

Problem 5-133: For the circuit below, find the AC input impedance between node V_{IN} and ground using the SPICE AC Analysis.

Problem 5-134: For the circuit below, find the AC output impedance between node V_{OUT} and ground using the SPICE AC Analysis.

Problem 5-135: For the circuit below, find the AC output impedance between node V_{OUT} and ground using the SPICE AC Analysis.

Problem 5-136: For the circuit below, find the AC output impedance between node V_{OUT} and ground using the SPICE AC Analysis.

Problem 5-137: For the circuit below, find the AC input impedance between node V_{IN} and ground using the SPICE AC Analysis.

PART 6
Time Domain Analyses

The time domain analyses are used to look at waveforms versus time. Multisim provides two methods of running time domain analyses, the virtual lab simulation using an Oscilloscope instrument, and the standard SPICE Transient Analysis. With these analyses, waveforms are displayed similar to how we would see them on an oscilloscope screen in the laboratory. With the Oscilloscope instrument, the waveforms are displayed almost exactly as you would see them in the lab. You have to set the oscilloscope settings to properly see the oscilloscope traces, and the waveform will dance around the oscilloscope screen if you do not properly trigger the scope. With the Oscilloscope instrument, it is easy to see a few cycles of a periodic waveform on the screen, and you usually do not view the complete waveform from time equals zero. With the SPICE Transient Analysis, you can look at the entire waveform or zoom in to a portion of the waveform. You do not need to set any parameters analogous to the oscilloscope settings to view the results of a Transient Analysis. Example waveforms versus time using the Transient Analysis and the Oscilloscope instrument are:

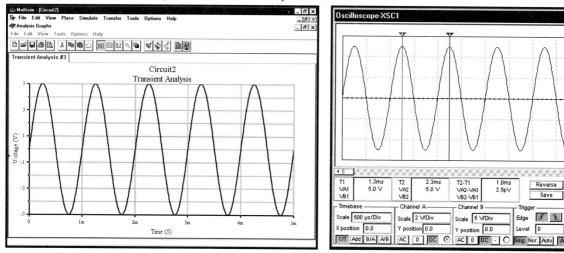

The graphs show us a voltage versus time. Use the Transient Analysis to obtain a waveform versus time. The time domain equation for this waveform is $v_x(t) = 5\sin(2\pi \bullet 1000t + 0°)$. This waveform has an amplitude of 5 V and a frequency of 1000 Hz. If you want to look at the magnitude and phase of voltages and currents, use the AC Sweep or Bode Plotter instrument. The magnitude of the above waveform is 5 V and the phase is zero degrees—in phasor notation, $5\angle 0°$. The AC Sweep or Bode Plotter instrument will give us the result $5\angle 0°$. The Transient Analysis and Oscilloscope instrument will give us the graphs above.

6.A. Using the Oscilloscope Instrument

The settings on the Oscilloscope instrument are similar to what you would see on any analog oscilloscope you would use in the lab. Its main settings are the timebase or time per division for the horizontal axis, volts per division for the two inputs that are displayed on the vertical axis, and the trigger setting. If you know how to use an analog oscilloscope, then you already know how to use these settings. (Note that there is no "Auto-Set" or "Auto-Scale" button for the Multisim Oscilloscope instrument.) If you are unfamiliar with using oscilloscopes, then using the Oscilloscope instrument in Multisim will give you the skills you need to use an oscilloscope in the lab.

We will start with a simple circuit and show how to display a waveform. Wire the circuit below:

ial, its voltage is not affected by the resistor. The resistor just absorbs power. The amplitude of the source is 1 volt, and its
This circuit uses the AC voltage source and a virtual resistor. The value of the resistor is not important. Since the source is
frequency is 10 kHz. To change the properties of the source, double-click the **LEFT** mouse button on the source graphic:

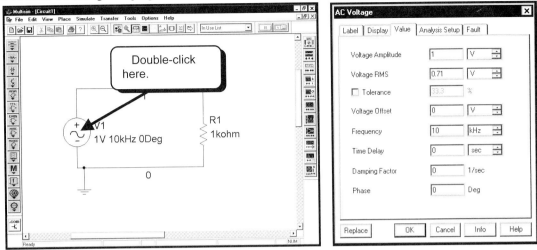

Set the parameters of your source to match those shown above. Also, make sure that your circuit is grounded as is the circuit shown above.

Next, we must place the Oscilloscope instrument in your circuit. Click the **LEFT** mouse button on the Oscilloscope button as shown. The Oscilloscope instrument will become attached to your mouse:

Place the Oscilloscope instrument in a convenient location:

We will look closely at the input terminals of the oscilloscope:

A, **B**, **G**, and **T** are terminals on the oscilloscope to which we can make connections. The ground terminal (**G**) allows you to connect the oscilloscope to ground. Most oscilloscopes are grounded through the three-terminal plug that you plug into the wall outlet. All oscilloscope probes have a measuring pin and a ground clip. Since the oscilloscope is grounded through the three-terminal plug, and a standard oscilloscope measures all voltage relative to ground, you do not have to use the ground clip on the oscilloscope probe to measure a voltage. However, experience will show you that using the ground clip on the oscilloscope probe greatly reduces the amount of noise you measure with the scope. Most oscilloscopes also have a ground lug somewhere on the front of the scope. This ground lug is usually a knurled knob that you can unscrew and attach a wire to, or a terminal to which you can connect a banana plug. This terminal provides you with another way to ground the oscilloscope to your circuit. Usually you will not use this terminal since the oscilloscope is already grounded through the three-terminal plug. Thus, it does not matter if you use the ground terminal on the front of the oscilloscope. The **G** terminal on the Multisim oscilloscope is analogous to the ground terminal on the front of your oscilloscope. If you do not hook up the **G** terminal, the Oscilloscope instrument will still be grounded, and will still measure voltages relative to ground. Thus, hooking up the **G** terminal to ground is optional, and does not affect the simulation of the circuit. Like oscilloscopes in the lab, the Multisim oscilloscope is automatically grounded.

The **A** and **B** terminals are the measurement inputs of the scope. These two terminals are connected to your circuit, and the voltages they measure are displayed on the oscilloscope screen. Lab oscilloscope inputs use 10:1 probes that have an input pin and a ground clip. The Multisim oscilloscope does not have the equivalent of a 10:1 probe, and there are no ground clips. On real oscilloscopes, the 10:1 probes with ground clips are necessary to reduce noise and allow the measurement of high-frequency waveforms. 10:1 probes and ground clips are not necessary in a simulation because all we are really doing with the Oscilloscope instrument is plotting the numerical data from a simulation. Thus, the Oscilloscope instrument only has a single connection for the **A** input and a single connection for the **B** input. Wiring either of these inputs to a node in the circuit causes the Oscilloscope instrument to display the voltage of that node relative to ground. The procedure is basically the same as you would do in the lab except that there is no 10:1 probe, and you do not need to connect the ground clip.

The **T** input is an external trigger input for the oscilloscope, and is analogous to the external trigger input of a real oscilloscope. Normally, oscilloscopes generate their trigger input from the waveform measured on either **Channel A** or **Channel B**. When measuring large waveforms, using the **A** or **B** input as the trigger input works well; however, if the measured waveforms on channel **A** or **B** are very small, the oscilloscope may have trouble generating a trigger from the small waveforms and the displayed waveforms will dance across the oscilloscope screen. One method to avoid this problem is to use the external trigger input (**T**) of the oscilloscope. The signal on the trigger could be a very large voltage waveform allowing the oscilloscope to easily generate a trigger, while the inputs on the **A** and **B** inputs are very small. The signal on the trigger input has to be synchronized to the waveforms being measured on the **A** and **B** inputs, but it can be much larger and allow the oscilloscope to easily generate a trigger. The external trigger input (**T**) on the Multisim Oscilloscope instrument has the same function as the external trigger input of a real scope. We will not use it in any of the simulations in this chapter because all of the waveforms we will measure are very large, and the oscilloscope can easily generate a trigger from either the **A** or **B** inputs.

We will first use **Channel A** to measure the voltage of the source. Wire the oscilloscope to your circuit as shown:

Remember that the oscilloscope is automatically grounded. To run the simulation, click the **LEFT** mouse button on the Run button as shown:

The simulation has started running, but we cannot see anything until we open the oscilloscope window. To open the oscilloscope window, double-click on the Oscilloscope instrument:

Your window may look similar to the one shown or not, depending on your oscilloscope's settings. After we change all of the settings, we should all see the same display.

6.A.1. Timebase

The first thing we will do is change the timebase. The present setting of the **Timebase** on my oscilloscope is **10 ms/Div**:

Note that the **Y/T** button is selected. This means that the horizontal (x) axis is the time axis, and all waveforms are displayed versus time. The y-axis is what is measured on **Channel A** and/or **Channel B**. The horizontal axis is time, and the **Scale** is **10 ms** per major division.

We are measuring a signal that has a 10 kHz frequency. The period of a signal is the reciprocal of the frequency, so the period of the measured signal is 0.1 ms, or 100 µs. The period of a signal is how long it takes the signal to repeat itself. Since our **Timebase** is set to 10 ms/div, and our measured signal repeats itself every 0.1 ms, there are 100 sine waves in every division on the horizontal axis. This is why the displayed waveform appears to be a solid thick black line.

Suppose we would like to display one sine wave of the measured waveform in one major division of the horizontal axis. (We would like one period to fit into one square.) One period is 0.1 ms or 100 µs, so we need to change the **Timebase** to 100 µs/div. To change the setting, click the *LEFT* mouse button in the **Scale** field. A cursor will be placed in the field, and up and down arrows will be displayed in the field:

You can use the mouse to click on the arrows and adjust the settings, or you can use the arrows on the keyboard to adjust the setting. Adjust the setting to 100 µs/div (not 100 **ms**/div). I will press the down arrow on the keyboard to change the setting:

EWB Multisim — Time Domain Analyses

You can change the **Timebase** as you see fit so that the waveform is displayed to your tastes. I will display two sine waves over the entire screen:

The **Add**, **B/A**, and **A/B** buttons will be discussed later.

6.A.2. Channel A and Channel B Volts per Division Settings

The volts per division setting works the same for both channels A and B. Since we are only measuring a single waveform on **Channel A**, we will confine our discussion to the **Channel A** controls. When you hook up the **Channel B** input, you can use the **Channel B** volts per division settings the same way as shown here.

The waveform we are measuring is a 1-volt-amplitude sine wave. This means that the signal's largest positive value is +1 volt, and its largest negative value is –1 volt. When we look at the **Channel A** volts per division setting, we see that it is set at **5 V/Div**:

The voltage measured on **Channel A** is displayed on the vertical or y-axis. With a **5 V/Div Scale**, each major division in the vertical direction represents 5 volts. Since our signal only has an amplitude of 1, the signal only occupies 1/5 of a division in the y-direction.

To make the measured waveform appear larger in the oscilloscope window, we need to change the **Channel A** volts per division setting. To have the waveform occupy one division in the positive direction and one division in the negative direction, we would set the **Channel A** volts per division setting to 1 V/div. To change the setting, click the *LEFT* mouse button in the **Scale** field. A cursor will be placed in the field, and up and down arrows will be displayed in the field:

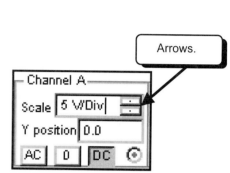

You can use the mouse to click on the arrows and adjust the settings, or you can use the arrows (⬆⬇) on the keyboard to adjust the setting. Adjust the setting to 1 V/div. I will press the down arrow ⬇ on the keyboard to change the setting:

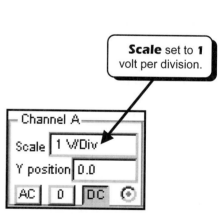

You can adjust the **Scale** setting to display the waveform to suit your tastes. The **Scale** setting below is 500 mV/div.

The **Y position** field is equivalent to the offset knob on a oscilloscope and it allows you to move the waveform up or down on the screen. For example, a **Y position** of 1.0 will shift the waveform up by 1 division:

A **Y position** of −1.0 will shift the waveform down by 1 division:

The offset knob on a real scope, and the **Y position** setting on the Multisim oscilloscope allow you to move an oscilloscope trace around the screen and allow you to display the waveforms the way that you think shows the waveforms best. For a single waveform, the **Y position** is usually set to zero. When you use **Channel A** and **Channel B**, you may want to separate the two traces so that they do not overlap, and you can use the **Y position** settings to move the traces around the oscilloscope window and display them to your liking.

The **AC**, **0**, and **DC** buttons are the coupling settings for the oscilloscope channel:

They are the same as the AC, ground, and DC coupling settings on a real oscilloscope. Setting the channel coupling to **0** (or ground on a real oscilloscope) sets the input of that channel to zero, even though that channel is connected to a node and measuring a voltage. With coupling set to **0**, the oscilloscope channel would then measure and display a constant value of zero volts. The **0** setting is used so that you can identify where the zero voltage reference is on your oscilloscope display, and you can then use the **Y position** to move the zero reference if you want.

The **DC** coupling setting displays everything contained in a waveform. If you are measuring a constant or DC voltage, the oscilloscope will display it. If you are measuring a pure AC waveform, such as the one we are measuring in this example, the oscilloscope will display it. If you are measuring a voltage that has an AC waveform with a DC offset, the oscilloscope will display both the waveform and the offset. Basically, the DC setting shows you everything: both the constant part and the time-varying part.

The **AC** coupling setting filters off the constant portions of a signal and only displays the AC portion of a waveform. If you are measuring a constant or DC voltage, the oscilloscope will display it as zero. If you are measuring a voltage that has an AC waveform with a DC offset, the oscilloscope will remove the DC part and display only the AC part. Basically, the **AC** setting shows you only the time-varying part of the waveform.

To illustrate the differences between the coupling settings, we will add a DC source to our circuit:

Since the two sources are in series, the sources add. Thus, the waveform being measured by the oscilloscope is 1 V DC plus a 10 kHz sine wave with an amplitude of 1 volt. We will display this waveform on the scope. First, we will set the oscilloscope coupling to **0**. This will show us where the 0 reference is for the oscilloscope display:

It is difficult to see, but the zero reference is now right in the center of the oscilloscope screen. You can change the **Y position** field to see the constant line shift up and down. When the **Y position** field is set to **0.0**, as in the screen capture above, the constant line will be right in the center of the oscilloscope display.

If we change the coupling to **DC**, a sine wave will be displayed. It will be centered on 1 volt rather than zero. Note that the volts per division setting is set at **1 V/Div** for the display below. To have the oscilloscope display correctly, you may need to set the trigger **Level** to **1** volt and the trigger mode to **Nor**mal. We see that when the oscilloscope is set to **DC** coupling, the oscilloscope displays both the DC offset and the AC waveform:

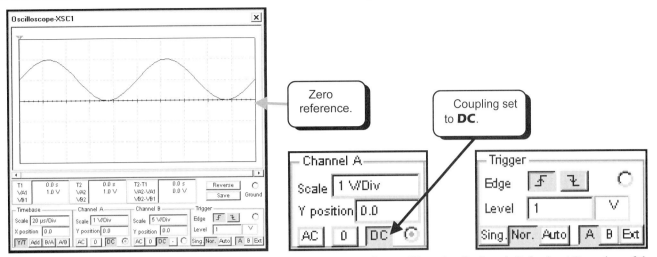

If we switch the oscilloscope coupling to **AC**, the DC part of the waveform will not be displayed. Only the AC portion of the waveform will be displayed, and it will be centered on the zero reference of the oscilloscope. To have the oscilloscope display correctly, you may need to set the trigger **Level** to 0 volt and the trigger mode to **Nor**mal. Here we see that AC coupling removes the DC offset and shows us only the time-varying (AC) part:

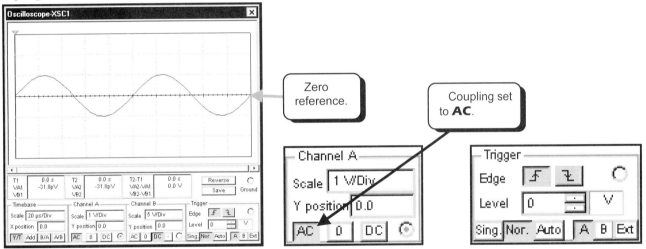

We see that DC coupling shows us both the AC and DC parts of a waveform, while the AC coupling only shows us the AC part and removes the constant or DC part.

As a last example of coupling, we will measure the voltage of the DC source using AC and DC coupling. Rewire the circuit as shown:

Next, simulate the circuit and set the coupling to DC. You will need to set the trigger mode to **Auto** to see anything on the oscilloscope screen:

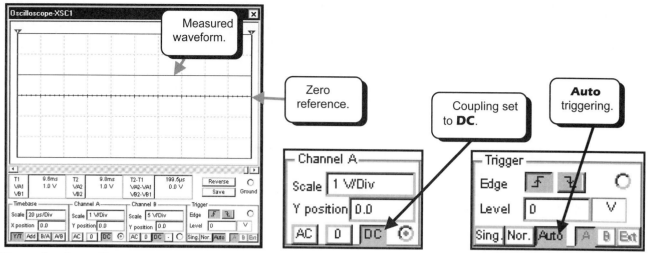

We see that the oscilloscope measures a constant voltage of 1 volt. If we switch the coupling to AC, the oscilloscope ignores the DC part of the waveform, and displays what is left. In this case, the waveform we are measuring is pure DC, so nothing is left, and the oscilloscope displays the measurement as zero:

The coupling setting you use depends on what you are attempting to measure. If the offset (DC portion) of a waveform is important, you should use DC coupling. If all you are interested in is the time-varying portion of a waveform, and you do not care what the DC part is, you can use AC coupling. DC coupling shows you everything, while AC coupling only shows you the time-varying part.

6.A.3. Trigger Settings

Triggering is one of the most difficult skills to master with a real oscilloscope. With an oscilloscope that uses a cathode ray tube (CRT), the trace is drawn on the screen by an electron beam that impacts the CRT screen. As electrons impact the screen, the phosphors on the screen glow and are seen by the user. You can think of the electron beam as a pen that writes on the screen. The pen starts at the left side of the screen and moves to the right. The **Timebase** setting tells the oscilloscope how fast to move the trace from the left to the right. The **Channel A** input tells the pen how to move up and down in the vertical direction. The trigger tells the pen when to start moving.

When drawing a line on the oscilloscope screen, the following steps happen. The pen stays on the left side of the screen and does not move until it receives a trigger. When a trigger occurs, the pen starts moving to the right and draws a visible line on the screen. When the pen reaches the right side, it stops drawing a trace and positions the pen on the left side of the screen. The pen remains stationary until it receives another trigger signal.

Thus, the trigger tells the oscilloscope when to start drawing the trace and it is used to synchronize the trace sweep with the measured waveform. If the oscilloscope cannot generate a trigger, the pen will never move, and a trace will not be drawn on the screen. In this case, a trace will not be displayed and the oscilloscope screen will be blank. If the oscilloscope randomly generates trigger signals that are not synchronized with the measured waveform, the measured waveform will be displayed, but it will dance on the screen and never appear to stand still. When an oscilloscope is properly triggered, a waveform will be displayed and it will appear still on the screen.

For all of the trigger modes discussed here, the oscilloscope can use the waveform on **Channel A**, **Channel B**, or the External Trigger input. You select the source with the buttons shown below:

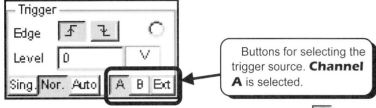

In the screen capture, the waveform on **Channel A** is selected. The **Edge** is set to positive going and the trigger **Level** is set to **0** volts. The oscilloscope generates a trigger by comparing the waveform of **Channel A** to the trigger level. When the waveform crosses the level going positive, a trigger is generated. In our example screen capture above, when the waveform on **Channel A** crosses 0 volts going positive, the pen will start moving from right to left and draw a trace on the screen. The resulting waveform is displayed for the circuit below. Note in the circuit that we have changed the amplitude of our source to 3 V for illustration purposes. Note also that for these examples, **Nor**mal triggering is selected:

We see that the trace starts when the waveform crosses zero in the positive going direction.

Next, leave all of the settings the same but change the trigger **Level** to **1 V**. This will cause the oscilloscope to trigger when the waveform is positive going and crosses 1 volt:

We see that the trace starts when the waveform crosses 1 volt in the positive going direction. Next, leave all of the settings the same but change the trigger **Level** to **2 V** and set the **Edge** to negative. This will cause the oscilloscope to trigger when the waveform is negative going and crosses 2 volts:

Note that the trace starts when the waveform crosses positive 2 volts. If we wanted the trace to start when the waveform is negative, we would need to set the level to a negative value, –2 volts for example:

As a last example, we will set the trigger level to 4 volts. Since the waveform never crosses 4 volts, a trigger will never be generated and a trace will not be displayed on the screen:

No trace is shown. Since we now understand oscilloscope triggering, we know that we need to decrease the level so that the waveform on **Channel A** crosses the **Level** setting.

The Multisim oscilloscope has three trigger modes, Single, Normal, and Auto:

Normal Triggering (Nor.) – All of the examples above used normal triggering. In this mode, the pen waits at the left side of the oscilloscope window for a trigger. A trigger is generated when the specified trigger source (**Channel A**, **Channel B**, or the External Trigger input) crosses the **Level**. At the trigger, the pen moves to the right and draws a trace on the screen. When the pen reaches the right side, it stops drawing the trace and then positions itself back on the left side of the window. It remains stationary until the next trigger is generated. With this mode, a measured waveform will appear stationary on the oscilloscope window.

Single Trigger (Sing) – Single trigger mode works the same as Normal trigger mode except that only a single trace is drawn across the screen. When you press the Sing button, the oscilloscope becomes armed and waits for a trigger. When a trigger occurs, a single trace is drawn on the screen, and then the trigger becomes disarmed. No more traces are drawn until the oscilloscope is once again armed by the user and a trigger is generated. This mode is usually used in situations where waveforms are not repetitive, such as a pulse that occurs only once.

If you have a conventional analog oscilloscope that has a single trigger mode, you would see a single trace drawn on the screen, and then the trace would slowly fade away as the phosphors on the screen slowly stopped glowing. A digital oscilloscope in single trigger mode knows that you want to measure a single event. After it records the single trace, it saves the waveform and continues to display it. The waveform is not erased until you arm the oscilloscope and record another single trace, or you switch the mode of the oscilloscope. Before the advent of digital scopes, there were analog oscilloscopes called "storage oscilloscopes." These oscilloscopes would allow you to measure a waveform in single trigger mode, but would continue to display the waveform long after the trigger occurred. In a conventional analog oscilloscope, the trace would slowly fade. With a storage oscilloscope, the single trace waveform was displayed on the screen as long as you needed it. With conventional analog scopes, digital scopes, and analog storage scopes, the single trigger mode works the same. The oscilloscopes just differ on how the trace is displayed and stored.

Auto Triggering (Auto) – With auto triggering, a trigger is not generated by comparing one of the inputs to the trigger. Instead, the trigger is generated internally. On many scopes, auto triggering uses the 60 Hz signal from the power line to generate a trigger. No matter how the trigger is generated, it is not synchronized to the waveform you are measuring. This will cause a measured waveform to move on the oscilloscope. The waveform you are measuring has nothing to do with the trigger. Thus, the point on your waveform where the trace starts is random and changes. As an example, if we use our previous circuit and switch the trigger mode to **Auto** Auto, a waveform will be displayed, but the starting point of the waveform will not be constant. Three screen captures are shown below to illustrate this point:

Auto trigger is usually used in two situations. The first is when you are measuring DC voltages. A DC voltage never changes. Therefore, the measured signal can never cross the level and a trigger is never generated if you are using single trigger mode or normal trigger mode. The only way to display a DC waveform is to switch to **Auto** triggering and have the trigger generated internally.

The second situation in which you would use **Auto** triggering is if you are having trouble getting the oscilloscope to trigger. If you cannot get the oscilloscope to generate anything in single or normal mode, switch the oscilloscope to Auto trigger mode. This will allow you to display something on the oscilloscope. Once you see some waveforms, you can measure the waveforms and see how to set the level to the appropriate value to get the oscilloscope to trigger.

We will not stress triggering in future examples. However, when using an oscilloscope, you must set the oscilloscope trigger properly or you will not see any waveforms on the oscilloscope screen.

6.A.4. Using the Cursors

The oscilloscope provides two cursors to allow us to make measurements of waveforms displayed on the oscilloscope window. We will continue with the previous circuit:

Simulate the circuit and set the **Timebase** to **50 μs/Div**. You may see one of the screens shown below:

My screen shows two cursors: cursor 1 is shown in red and cursor 2 is shown in blue. Your cursors may not be as easily seen as those above. It may be that both cursors are located at the leftmost portion of the oscilloscope window as shown:

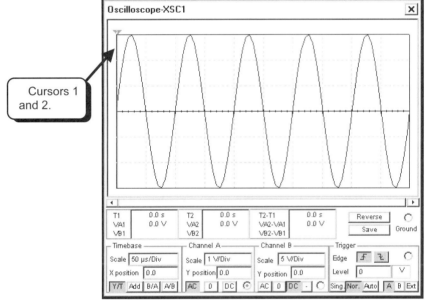

Since the cursors are in the same location, they are displayed in green. We can easily separate the cursors if they are positioned as shown above. Click **and hold** the *LEFT* mouse button on the green triangle as shown below:

While continuing to hold down the mouse button, drag the mouse. Cursor 1 will move with the mouse. When you release the mouse button, cursor 1 will be repositioned:

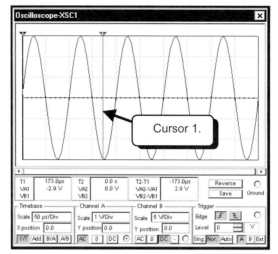

The cursors are now separated and easily identified.

We can now position each cursor separately using the method shown above. To move cursor 1, drag the red triangle. To move cursor 2, drag the blue triangle. As a first measurement, we will measure the period of this waveform. Position the cursors at two of the zero crossings:

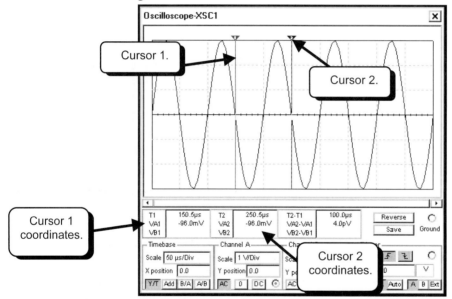

The cursor information is displayed in the text window below the oscilloscope screen. The coordinates of each trace are displayed for each cursor:

T1 is the time coordinate of cursor 1. **VA1** is the voltage of **Channel A** at cursor 1. **VB1** is the voltage of **Channel B** at cursor 1. Since there is no input on **Channel B**, **VB1** is always zero for this example. **T2** is the time coordinate of cursor 2. **VA2** is the voltage of **Channel A** at cursor 2. **VB2** is the voltage of **Channel B** at cursor 2. Since there is no input on **Channel B**, **VB2** is always zero for this example. The rightmost portion of the cursor information is the difference between the two cursors. We see that the time difference between the two cursors is 100 μs. This is the period of the waveform and corresponds to a frequency of 10 kHz.

We will next measure the peak-to-peak swing of the waveform. Position the cursors as shown:

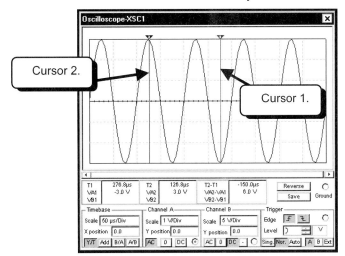

An enlarged view of the cursor information is:

We see that the difference between the voltages of cursor 2 and cursor 1 is **6.0** volts. This is the peak-to-peak voltage of the waveform. The peak-to-peak value is twice the amplitude, so this measurement agrees with the 3-volt amplitude specified in the schematic.

6.B. Phase Measurements of a Capacitive Circuit

For a simple example, we would like to measure the magnitude and phase of the capacitor voltage in the circuit below:

Before we measure the circuit with the oscilloscope, we will make some calculations. The impedance of the capacitor at 1000 Hz is

$$Z_C = \frac{1}{j\omega C} = \frac{1}{j2\pi(1000\,\text{Hz})(10^{-6}\,\text{F})} = -j159.155\,\Omega = 159.155\angle -90°.$$

We will use the input voltage as our reference, so the phase of the input is taken as zero degrees, $V_{IN} = 10\angle 0°$ volts. Using a voltage divider, we can obtain the capacitor voltage:

$$V_C = V_{IN}\left(\frac{Z_C}{Z_C + R}\right) = (10\angle 0°\,\text{volts})\left(\frac{-j159.155\,\Omega}{-j159.155\,\Omega + 100\,\Omega}\right) = (7.17 - j4.505)\,\text{volts} = 8.467\angle -32.142°\,\text{volts}$$

From these examples, we expect the magnitude of the capacitor voltage to be about 8.5 volts and the phase of the capacitor voltage to lag the input voltage by about 32 degrees.

We will first measure the magnitude of the capacitor voltage. Add the Oscilloscope instrument as shown:

We are measuring the input voltage with **Channel A** and the capacitor voltage with **Channel B**. Click on the Run button ![run] to start the simulation and then double-click on the Oscilloscope instrument to display the oscilloscope window:

It may not be obvious which trace is which on the oscilloscope. However, from our calculations we know that the capacitor voltage is smaller than the input voltage, so we can identify the traces by their size.

Next, we will use the cursors to measure the magnitude of the capacitor voltage. See section 6.A.4 for instructions on using the oscilloscope cursors. Place the cursors as shown:

The cursors are located at the peaks of the capacitor voltage waveforms. The measurements displayed by the cursors are:

We are measuring the capacitor voltage with **Channel B**, so the VB measurements correspond to the capacitor voltage. We see that the capacitor peak-to-peak voltage is 16.9 volts. The peak-to-peak value is twice the amplitude, so a peak-to-peak measurement of 16.9 volts corresponds to an amplitude of 8.45 volts. The calculated value of the capacitor voltage magnitude was 8.467 volts. Thus, our measurement agrees with our calculation.

Next, we will measure the phase of the capacitor voltage relative to the phase of the input voltage. We will first look at our present oscilloscope display and identify which trace is leading:

We identify the capacitor voltage as the one with the smaller amplitude. From the screen above, we conclude that the input voltage is leading the capacitor voltage, or that the capacitor voltage is lagging the input voltage. When we calculate the phase angle of the capacitor voltage, we will write it as a negative number because the capacitor voltage is lagging the input voltage.

To measure the phase angle, we need to measure the time difference between when the two traces cross zero. To get an accurate measurement, we need to blow up the time axis. Change the **Timebase** to **20 μs/Div** and position the cursors at the zero crossing points:

The difference in time is used to calculate the phase angle. The cursor readings are shown below:

We see that the time difference between the cursors is Δt = **90.3 μs**. We must now convert this time difference into a phase angle. One period or one cycle of a sine wave corresponds to 360°. Since we know the period of our signal, we can use the relationship:

$$\frac{\Delta t}{T} = \frac{\theta}{360°}$$

where T is the period in seconds and θ is the phase angle in degrees. Our source has a frequency of 1 kHz, so the period of our waveforms is 1 ms. We can easily solve this equation for θ:

$$\theta = 360° \frac{\Delta t}{T} = 360° \frac{90.3\,\mu s}{1\,ms} = 32.4°$$

Earlier we found that the capacitor voltage lags the input voltage, so the phase angle is negative. Thus, our measured phase angle is –32.4°. Our calculated value was –32.14°, so once again the measured result agrees well with the calculated result.

EXERCISE 6-1: Simulate the RC circuit below and find the magnitude and phase of the capacitor voltage.

ANSWER: The calculated value of the capacitor voltage is: $8.467 \angle -32.142°$ volts.

EXERCISE 6-2: Simulate the RC circuit below and find the magnitude and phase of the capacitor voltage.

ANSWER: The calculated value of the capacitor voltage is: $8.467 \angle -32.142°$ volts.

6.C. Phase Measurements of an Inductive Circuit

In this section we will look at the magnitude and phase response of an inductive circuit. From a simulation point of view, observing the magnitude and phase of an inductive circuit is no different than what we did in the previous circuit. To make this simulation a little different, we will display the inductor current rather than the inductor voltage. To display the current we need a sensor to sense the current and create a voltage signal that corresponds to the measured current. Multisim recommends using a current-controlled voltage source with a gain of 1. That is, this device senses a current, and the output voltage is 1 times the measured current. An example circuit that uses this method is:

Source **V2** is a current-controlled voltage source. The output of V2 is 1 Ω times the sensed current, which is the inductor current in this case. The gain of the sensor is 1 Ω. The units must be ohms because the output of the sensor is the gain times the sensed current, which has units of amps. For the output have to units of volts and the sensed quantity to have units of amps, the gain must have units of ohms. **R2** was added to prevent the error of having only one element connected to a node. The value of R2 does not affect anything and is arbitrary.

The problem with this method is that in the lab you probably do not have a current sensor like the one shown, and thus this virtual simulation would not closely match how you would do this in the lab. Several current sensors are available, but most are expensive. We will use a very inexpensive sensor called a current sensing resistor. An example circuit is:

Resistor **Rsense** was added to the circuit and is a current sensing resistor. The value of a current sensing resistor is small so that adding it to the circuit does not greatly affect the circuit's operation. However, the current through **Rsense** is the same as the current you want to measure. To measure the current, you measure the voltage across the sensing resistor and then divide by the resistance.

The value of the current sensing resistor depends on two factors. (1) It must be small compared to other impedances in the circuit so that it does not affect the current you are trying to measure. (2) It must not be so small that the measured voltage is tiny and hard to measure with your instruments. Thus, the value of a current sensing resistor is a compromise between the effect the resistor has on the circuit and the size of the voltage produced by the resistor. In our example circuit, **Rsense** is 10 Ω and R is 1 kΩ. Thus **Rsense** introduces an error of 1% in our measurements. The current we measure will be on the order of 10 mA, so the voltage across **Rsense** will be on the order of 100 mV. This voltage is easily measured with an oscilloscope.

We will now calculate the magnitude and phase of the inductor current. We will ignore the current sensing resistor, because we are interested in the original circuit. **Rsense** was added in the lab as a method to measure the current. We will calculate the current for the circuit below:

The impedance of the inductor at 10 kHz is

$$Z_L = j\omega L = j2\pi(10000\,\text{Hz})(0.01\,\text{H}) = j628.319\,\Omega = 159.155\angle 90°$$

We will use the input voltage as our reference, so the phase of the input is taken as zero degrees, $V_{IN} = 10\angle 0°$ volts. Since this is a series circuit:

$$I_L = \frac{V_{IN}}{Z_L + R} = \frac{10\angle 0°\,\text{volts}}{1000\,\Omega + j628.319\,\Omega} = (7.17 - j4.505)\,\text{mA} = 8.467\angle -32.142°\,\text{mA}$$

From these examples, we expect the magnitude of the current to be about 8.5 mA and the phase of the current to lag the input voltage by about 32 degrees.

We will first measure the magnitude of the current. Add the Oscilloscope instrument and current sensing resistor as shown:

We are measuring the input voltage with **Channel A** and the current sensing resistor voltage with **Channel B**. Click on the Run button to start the simulation and then double-click on the Oscilloscope instrument to display the oscilloscope window:

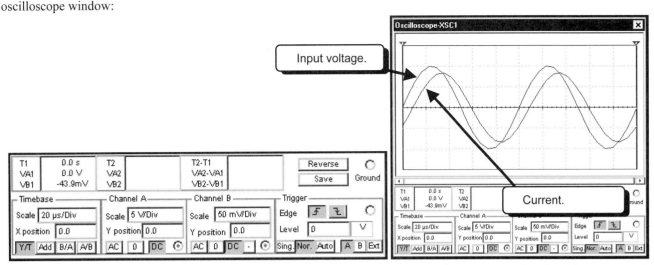

The two traces are easy to identify when you change the volts per division setting for **Channel B**. The voltage across the current sensing resistor is very small. As you change the **Channel B** volts per division setting, you can see which trace changes.

Next, we will use the cursors to measure the magnitude of the current. See section 6.A.4 for instructions on using the oscilloscope cursors. Place the cursors as shown:

The cursors are located at the peaks of the current waveform. The measurements displayed by the cursors are:

We are measuring the current sensing resistor's voltage with **Channel B**, so the VB measurements correspond to the current. We see that the resistor's peak-to-peak voltage is 166.7 mV. The peak-to-peak value is twice the amplitude, so a peak-to-peak measurement of 166.7 mV corresponds to an amplitude of 83.4 mV. To find the current, we need to divide the voltage across the current sensing resistor by its resistance:

$$Current = \frac{V_{RSense}}{R_{Sense}} = \frac{83.4 \, \text{mV}}{10 \, \Omega} = 8.34 \, \text{mA}$$

The calculated value of the magnitude of the current was 8.467 mA. The measured value is within 1.5% of the calculated value. A slight error is expected because we neglected the current sensing resistor in our calculations and because the current signal is a fairly small signal.

Next, we will measure the phase of the current relative to the phase of the input voltage. We will first look at our present oscilloscope display and identify which trace is leading:

From the screen above, we conclude that the input voltage is leading the current, or that the current is lagging the input voltage. When we calculate the phase angle of the current, we will write it as a negative number because the current is lagging the input voltage.

To calculate the phase angle, we need to measure the time difference between when the two traces cross zero. To get an accurate measurement, we need to blow up the time axis. Change the **Timebase** to **2 μs/Div** and position the cursors at the zero crossing points:

The difference in time is used to calculate the phase angle. The cursor readings are shown below:

T1	0.0 s	Time difference between the cursors.	T2-T1	9.0μs
VA1	0.0 V		VA2-VA1	5.3 V
VB1	-43.9mV		VB2-VB1	44.2mV

We see that the time difference between the cursors is Δt = **9.0 μs**. We must now convert this time difference into a phase angle. One period or one cycle of a sine wave corresponds to 360°. Since we know the period of our signal, we can use the relationship:

$$\frac{\Delta t}{T} = \frac{\theta}{360°}$$

where T is the period in seconds and θ is the phase angle in degrees. Our source has a frequency of 10 kHz, so the period of our waveforms is 0.1 ms. We can easily solve this equation for θ:

$$\theta = 360° \frac{\Delta t}{T} = 360° \frac{9.0\,\mu s}{0.1\,ms} = 32.4°$$

Earlier we found that the current lags the input voltage, so the phase angle is negative. Thus, our measured phase angle is −32.4°. Our calculated value was −32.14°, so once again the measured result agrees well with the calculated result.

EXERCISE 6-3: Simulate the LR circuit of this section, but use a current-controlled voltage source with a gain of 10 Ω rather than a current sensing resistor. Show that the results are nearly the same as we saw in this section and that measured values agree with the calculated values. The current-controlled voltage source is located in the same toolbar as the AC voltage source.

SOLUTION: Simulate the circuit below:

EXERCISE 6-4: Simulate the RC circuit below and find the magnitude and phase of the capacitor current.

ANSWER: The calculated value of the capacitor current is: $0.0532 \angle 57.86°$ amps.

6.D. Series LCR Resonant Circuit

As a last example of magnitude and phase measurements, we will look at a series RLC circuit at the resonant frequency:

Note that we have labeled the nodes as **V1**, **V2**, and **V3**. This was done so that we can easily talk about the nodes. You do not need to label the nodes if you do not want to. For this example, we are interested in the input voltage, the resistor voltage, the capacitor voltage, and the inductor voltage. All oscilloscopes measure voltages relative to ground. Thus, we can easily measure the voltages V1, V2, and V3 with an oscilloscope. In this circuit, the input voltage and the inductor voltage are easy to measure because the input voltage is equal to V1 and the inductor voltage is equal to V3. Measuring the resistor voltage and the capacitor voltage is a little more difficult. The resistor voltage is equal to V1 – V2, and the capacitor voltage is equal to V2 – V3. Most oscilloscopes provide a subtraction function so that you can display one **Channel** minus the other, **Channel A** minus **Channel B** for example. These oscilloscopes could display the resistor or capacitor voltage. The problem is that with some scopes, when you display **Channel A** minus **Channel B**, you cannot display anything else. Thus you could display the resistor voltage, but you may not be able to compare it to the input voltage because the oscilloscope can only display the subtraction by itself. This limitation depends greatly on your equipment. With an advanced four-channel oscilloscope, you could display all voltages of interest at the same time.

The Multisim oscilloscope is limited in that it only displays a single trace when it displays **Channel A** minus **Channel B**. To display the resistor and capacitor voltages relative to other voltages in the circuit, we will use a trick to reference these voltages to ground. This will be shown later.

Before we measure the circuit with the oscilloscope, we will make some calculations. We are interested in measuring the voltages at the resonant frequency, so the first thing we need to do is calculate the resonant frequency ω_o:

$$\omega_o = \frac{1}{2\pi\sqrt{LC}} = \frac{1}{2\pi\sqrt{(0.01\,\text{H})(1\times 10^{-6}\,\text{F})}} = 1592\,\text{Hz}$$

We can now calculate the impedance of the capacitor and inductor at resonance:

$$Z_C = \frac{1}{j\omega C} = \frac{1}{j2\pi(1592\,\text{Hz})(1\times 10^{-6}\,\text{F})} = -j100\,\Omega = 100\angle -90°$$

$$Z_L = j\omega L = j2\pi(1592\,\text{Hz})(0.01\,\text{H}) = j100\,\Omega = 100\angle 90°$$

Note that at resonance, the impedance of the inductor and the impedance of the capacitor add to zero.

We will use the input voltage as our reference, so the phase of the input is taken as zero degrees, $V_{IN} = 10\angle 0°$ volts. Using a voltage divider, we can obtain the voltage of all the elements in the circuit:

$$V_R = V_{IN}\left(\frac{R}{Z_C + Z_L + R}\right) = (10\angle 0° \text{ volts})\left(\frac{10\,\Omega}{-j100\,\Omega + j100\,\Omega + 10\,\Omega}\right) = 10 \text{ volts} = 10\angle 0° \text{ volts}$$

$$V_C = V_{IN}\left(\frac{Z_C}{Z_C + Z_L + R}\right) = (10\angle 0° \text{ volts})\left(\frac{-j100\,\Omega}{-j100\,\Omega + j100\,\Omega + 10\,\Omega}\right) = -j100 \text{ volts} = 100\angle -90° \text{ volts}$$

$$V_L = V_{IN}\left(\frac{Z_L}{Z_C + Z_L + R}\right) = (10\angle 0° \text{ volts})\left(\frac{j100\,\Omega}{-j100\,\Omega + j100\,\Omega + 10\,\Omega}\right) = j100 \text{ volts} = 100\angle 90° \text{ volts}$$

We see that the resistor voltage is equal to the input voltage (same phase and magnitude). The capacitor voltage lags the input voltage by 90°, and the inductor leads the input voltage by 90°. The inductor and capacitor voltages are equal in magnitude and 180 degrees out of phase. Thus, the inductor voltage and the capacitor voltage add to zero.

We will now modify the circuit to easily measure the resistor and capacitor voltages. We will add voltage-controlled voltage sources with a gain of one to the circuit:

An enlargement of a voltage-controlled voltage source is:

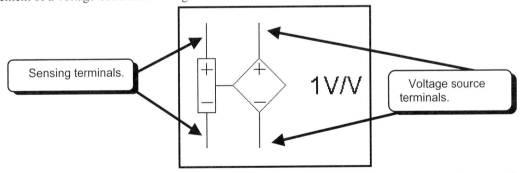

The sensing terminals can be hooked to any two nodes and measure the difference in voltage between those nodes. The voltage of the source is equal to the gain times the sensed voltage. Since we set the gain of our sources to 1, the output voltage of the source is equal to the measured voltage. In the circuit above, the sources measure the resistor voltage and the capacitor voltage. The negative terminals of the sources are grounded. This references the sources to ground, allowing the voltage of the sources to be measured with the oscilloscope. We have labeled the important nodes in the circuit as Vin, VR, VC, and VL because they are the voltages in which we are interested, and they can all be easily measured by an oscilloscope.

You might say that the method we are showing here is purely academic because we cannot use a voltage-controlled voltage source in the lab. This is true, but we can create an OPAMP circuit that implements the same function. We are basically using the voltage-controlled voltage source as a difference amplifier (an amplifier that subtracts one voltage from another). One method of implementing a difference amplifier is shown below:

If you want to, you can use this OPAMP circuit rather than the voltage-controlled voltage source. You will need to make two copies, one for the resistor voltage and one for the capacitor voltage. Using the circuit offers no advantage as far as the simulation is concerned, and it will take a long time to draw. The OPAMP circuit shown can only handle small voltages in the range of about –15 to +15 volts. You can use this circuit in the lab, but you will need additional circuitry to reduce the voltages seen by the circuit.

The voltage-controlled voltage source is located in the Sources toolbar. Click the **LEFT** mouse button on the **Sources** button :

The voltage-controlled voltage source is located as shown below:

When you click on the button, the source will become attached to the mouse pointer, and you can place the part.

We will first measure the input voltage and the resistor voltage. Add the Oscilloscope instrument as shown:

We are measuring the input voltage with **Channel A** and the resistor voltage with **Channel B**. Click on the Run button to start the simulation and then double-click on the Oscilloscope instrument to display the oscilloscope window:

Two traces are shown on the oscilloscope screen even though they look like a single trace. The two traces are nearly identical, so they appear as a single trace. If you change the volts per division or the **Y position** of one of the channels, you will see that two traces are displayed. I will change the **Scale** for **Channel B** to 10 V/Div:

We can now see that two traces are displayed. In our first oscilloscope capture, the settings for both channels were the same and the traces were identical, so they looked like a single trace on the oscilloscope screen. Our simulation shows that **VIN** and **VR** are equal within the measurement accuracy of the oscilloscope. This agrees with our calculations that **VIN** was equal to **VR**.

Next, we will measure the magnitude of the capacitor voltage and compare it to the input voltage. Connect **Channel B** of the oscilloscope to node **VC** and display the waveforms:

We are measuring the input voltage with **Channel A** and the capacitor voltage with **Channel B**. Click on the Run button to start the simulation and then double-click on the Oscilloscope instrument to display the oscilloscope window. Note in the screen capture below that I have not yet changed the **Scale** setting for **Channel B**:

The capacitor voltage is much larger than the resistor voltage of the previous measurement. The voltage is so large that it cannot be displayed with the **Scale** set to 10 V/Div. Change the **Scale** for **Channel B** to **50 V/Div**.

Input voltage.

Capacitor voltage.

A quick look at the oscilloscope screen shows that the two waveforms have about the same amplitude in terms of the number of divisions on the oscilloscope graticule, but the **Scale** setting for **Channel B** is ten times as large as the **Scale** setting for **Channel A**. We conclude that the capacitor voltage is about 10 times as large as the input voltage.

Next, we will use the cursors to measure the magnitude of the capacitor voltage. See section 6.A.4 for instructions on using the oscilloscope cursors. Place the cursors as shown:

Cursor 2 (blue).

Cursor 1 (red).

The cursors are located at the peaks of the capacitor voltage waveforms. The measurements displayed by the cursors are:

Peak-to-peak capacitor voltage.

We are measuring the capacitor voltage with **Channel B**, so the VB measurements correspond to the capacitor voltage. We see that the capacitor peak-to-peak voltage is 199.6 volts. The peak-to-peak value is twice the amplitude, so a peak-to-peak measurement of 199.6 volts corresponds to an amplitude of 99.8 volts. The calculated value of the capacitor voltage magnitude was 100 volts. Thus, our measurement agrees with our calculation.

Next, we will measure the phase of the capacitor voltage relative to the phase of the input voltage. From our calculations, we expect the capacitor voltage to be 90° out of phase with the input. Instead of using the technique presented in sections 6.B and 6.C, we remember that if two sine waves are 90° out of phase, one sine wave crosses zero when the other sine wave is at its peak. We can easily check this by lining up the cursors at one of the zero crossings:

It does appear that the zero crossings of one waveform line up with the peaks of the other waveform, so we conclude that the waveforms are approximately 90° out of phase.

Next, we will measure the magnitude of the inductor voltage and compare it to the input voltage. Connect **Channel B** of the oscilloscope to node **VL** and display the waveforms:

We are measuring the input voltage with **Channel A** and the inductor voltage with **Channel B**. Click on the Run button to start the simulation and then double-click on the Oscilloscope instrument to display the oscilloscope window:

The oscilloscope settings are the same as they were for the last capacitor measurement. The inductor voltage looks nearly the same as the capacitor voltage. If you identify the traces, you will see that the inductor voltage leads the input voltage, while in the previous example, the capacitor voltage lags the input voltage.

Next, we will use the cursors to measure the magnitude of the inductor voltage. See section 6.A.4 for instructions on using the oscilloscope cursors. Place the cursors as shown:

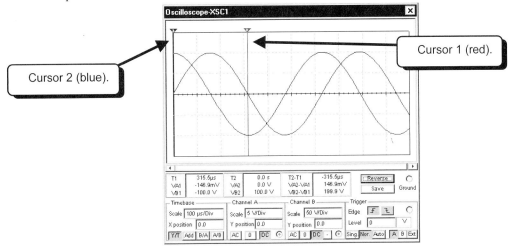

The cursors are located at the peaks of the inductor voltage waveforms. The measurements displayed by the cursors are:

We are measuring the capacitor voltage with **Channel B**, so the **VB** measurements correspond to the inductor voltage. We see that the inductor peak-to-peak voltage is 199.9 volts. The peak-to-peak value is twice the amplitude, so a peak-to-peak measurement of 199.9 volts corresponds to an amplitude of about 100 volts. The calculated value of the inductor voltage magnitude was 100 volts. Thus, our measurement agrees with our calculation.

Next, we will measure the phase of the inductor voltage relative to the phase of the input voltage. From our calculations, we expect the inductor voltage to be 90° out of phase with the input. Using the same technique we used for the capacitor, we verify that the two waveforms are 90° out of phase. In the previous screen captures, we note that the zero crossings of one waveform line up with the peaks of the other waveform, so we conclude that the waveforms are approximately 90° out of phase.

Next, we will measure the inductor voltage and the capacitor voltage. Change the circuit as shown so that you measure **VC** with **Channel A**:

Run the simulation and change the **Scale** for **Channel A** to **50 V/Div**:

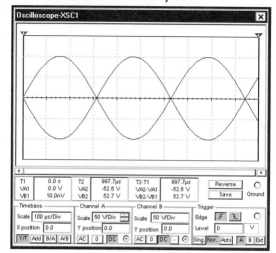

We see that the amplitudes of the waveforms are approximately equal, and that the two waveforms are 180° out of phase, which is the expected result.

For our last measurement, we will measure the input voltage with **Channel A** and the voltage across the capacitor and inductor with **Channel B**. I have connected **Channel B** to the node labeled **Vmid** in my circuit. You do not need to label the node in your circuit. You only have to connect **Channel B** to the same node in your circuit.

When you run the simulation, you will note that the voltage measured by **Channel B** is very small. I have changed the **Scale** on **Channel B** to **100 mV/Div**. The **Scale** setting for **Channel A** is **5 V/Div**:

The calculated voltage across the inductor and capacitor is supposed to be zero. In our simulation, we measure a peak-to-peak value of 446.4 mV, corresponding to an amplitude of 223 mV. The voltage is not exactly zero because our signal source is close to, but not exactly at the resonant frequency. The measured voltage is small enough that we conclude that the circuit is working properly, and that our calculations that the voltage goes to zero are also correct.

EXERCISE 6-5: Simulate the parallel RLC circuit below and find the magnitude and phase of the currents through each element, and the voltage across the resistor.

ANSWER: Calculated values: The resistor current is $1\angle 0°$ amps, the capacitor current is $0.1\angle 90°$ amps, the inductor current is $0.1\angle -90°$ amps, and the resistor voltage is $10\angle 0°$ volts.

6.E. Regulated DC Power Supply

We would like to simulate the operation of a voltage-regulated DC power supply. The complete circuit is shown below:

This is a 15 V DC power supply designed for a maximum output current of 1 A. The MDA2502 is a 25-amp, 200-volt bridge rectifier. This part has a much larger current specification than needed for this application, but it is a standard part provided with the Multisim program. This bridge is a standard part made by several manufacturers including International Rectifier (GBPC2502A), Diodes Incorporated (GBPC2502), and General Semiconductor (GBPC2502). The part number MDA2502 is a Motorola number, and is now obsolete.

Since this circuit contains three new parts, we will show how to find them. To place the transformer, click the **LEFT** mouse button on the **Basic** parts button, , to obtain the **Basic** parts toolbar:

Click the **LEFT** mouse button on the **Transformers** button to obtain a list of available transformers:

We would like to use a 24 V RMS transformer, so select the TS_PQ4_24:

Click on the **OK** button and place it in your circuit. This is a 24 V transformer, so we do not need to modify any of the settings once we place it in our circuit.

To place the bridge component, click the **LEFT** mouse button on the **Diodes** button :

Click the **LEFT** mouse button on the **FWB** (Full-Wave Bridge) button, to obtain the list of available bridges:

Select the **MDA2501**:

Click on the **OK** button to place the part in your circuit.

The last new part we will show how to place is the 7815 linear voltage regulator. This part can be found by using the **Analog** toolbar. Click the **LEFT** mouse button on the **Analog** button:

You can select any of the buttons in the left column of this toolbar:

A voltage regulator is not one of the standard component families that can be accessed using a button, so we must manually specify the **Component Family** for this group. Click the *LEFT* mouse button on the down triangle ▼ as shown:

Select **VOLTAGE_REGULATOR** from the list to display the available linear voltage regulators:

This list displays all of the standard linear voltage regulators available in Multisim. Select the **LM7815CT**:

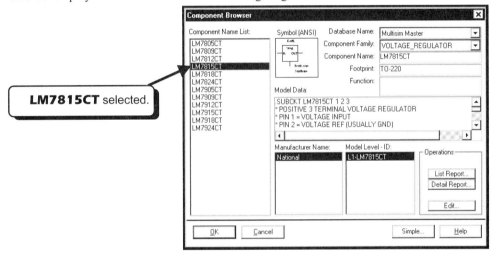

Click on the **OK** button to place the component in your circuit.

We can simulate this circuit with either the virtual lab simulation using an Oscilloscope instrument or the SPICE Transient Analysis. The virtual simulation gives us a feel for what we should see in the lab, while the Transient Analysis is a more powerful simulation that can give us information about every voltage and current in the circuit. We will show both here.

6.E.1. Regulated DC Power Supply – Virtual Lab Simulation

We will first run the virtual simulation. To add the Oscilloscope instrument to your circuit, click the **_LEFT_** mouse button on the Oscilloscope instrument button:

Place the oscilloscope in your circuit and wire it as shown:

We are now ready to run the simulation, so click on the **Run** button to start the simulation.

The simulation will start running, but we cannot see the results until we open the oscilloscope window. Double-click the *LEFT* mouse button on the oscilloscope to open its window:

Waveforms are shown, but they are off the screen. If your oscilloscope is set like mine, the **Channel A** and **Channel B** volts per division are set to 5 volts per division (**5 V/Div**). The controls are enlarged below:

Change the setting to **10 V/Div** for both **Channel A** and **Channel B**:

As you change the settings, the display of the waveforms on the oscilloscope screen will also change:

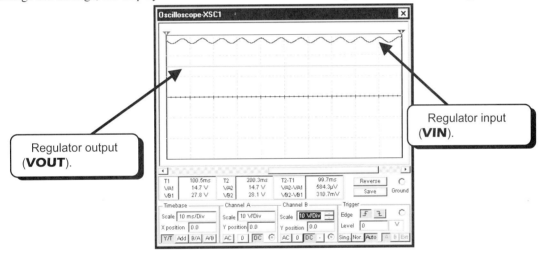

The output appears to be relatively constant at 15 V or so, and the input has a large amount of 60 Hz ripple. This is the function of a voltage regulator. As long as the input is large enough, the output of the regulator will be constant even though there is a large amount of ripple on the input.

If we want to see the ripple on the output of the regulator, we need to specify a small number of volts per division (say 10 mV/Div) for the output trace. With this setting, a 15 V DC signal would be way off the oscilloscope screen. To see the AC portion of the signal and filter off the DC portion we need to set the oscilloscope coupling to AC. Click the **LEFT** mouse button on the AC button for **Channel A** as shown below:

The trace will shift down to the center of the screen:

To see the ripple, we need to change the volts per division setting for **Channel A**. Change the setting to blow up the ripple so that it can easily be viewed on the oscilloscope screen. I chose 2 mV per division:

We see that there is ripple at the output, but it is very small when compared to the 15 V DC output.

To get a better idea of how the ripple on the input relates to the ripple on the output, we can change the **Timebase** setting using the same method we used to change the volts per division. I will change the setting to two **2 ms/Div**:

You should see the oscilloscope display below:

To see what the ripple on the input looks like, change the settings for **Channel B** to **AC** coupling and **1 V/Div**:

These settings show us that the ripple on the input and output has nearly the same shape:

Since the waveforms have the same amplitude, and the volts per division setting for **Channel B** is 500 times larger than the volts per division setting for **Channel A**, we conclude that the ripple at the output is 500 times smaller than the ripple at the input for this regulator.

The specification for the relationship between the ripple on the output as a function of the ripple on the input is called the ripple rejection ratio, and is defined as:

$$\text{Ripple Rejection Ratio}_{\text{dB}} = 20\log_{10}\left(\frac{\Delta Vin}{\Delta Vo}\right)$$

For our simulation, we found that the ratio $\frac{\Delta Vin}{\Delta Vo}$ was equal to 500. This corresponds to a Ripple Rejection Ratio of 54 dB.

The National Semiconductor data sheet for an LM7815C specifies the Ripple Rejection Ratio to be a minimum of 54 dB and a typical value of 70 dB. So we conclude that the Multisim model for the LM7815CT agrees with the minimum specifications for the National Semiconductor version of the LM7815C.

The last thing we might want to check on our circuit is the actual DC voltage at the output. We could use the oscilloscope cursors to find this value. Instead, we will use the Multimeter to measure the DC voltage at the output. Turn off the simulation by clicking on the **Run/STOP** button. Close the oscilloscope window and place the Multimeter instrument as shown. See section 3.A.2 for instructions on using the Multimeter instrument.

Start the simulation by clicking on the **Run** button. Double-click the **LEFT** mouse button on the Multimeter to open its window. Select the DC voltage setting. After a few moments the meter reading will become steady and display the DC voltage at the output. My meter reads a voltage of **14.732** volts:

The National Semiconductor datasheet for the LM7815C specifies the output voltage to be between 14.4 and 15.6 volts, so the output voltage is within range for this device.

EXERCISE 6-6: Simulate a power supply using a half-wave rectifier and a regulator. Use the same circuit as on page 327, except use a half-wave rectifier instead of a full-wave rectifier. Use the virtual diode because Multisim does not have a discrete diode that can handle the required current in the library. The remainder of the circuit should be the same.

SOLUTION:

Run the simulation and then display the results on the oscilloscope window. You will need to run the simulation for a long time because it takes the capacitor a long time to reach steady state. When you first start the simulation, you will see waveforms similar to the ones shown below for the input and output of the regulator:

As the simulation runs, the minimum peak of the capacitor voltage increases:

We see that the ripple on the output of the regulator becomes less because the input to the regulator is larger. Eventually, the capacitor charges up to a voltage high enough that the capacitor voltage is above the minimum voltage required by the regulator, and the output of the regulator is constant at 15 volts:

It appears that the circuit still functions properly since the output of the regulator is constant at 15 V. However, in this case, it took several cycles for the circuit to reach a constant output voltage. When you look at the input to the regulator, the ripple voltage across the capacitor is about twice the magnitude of the ripple voltage of the full-wave rectifier. This is the expected result.

6.E.2. Regulated DC Power Supply – SPICE Transient Simulation

The Multisim virtual simulation shows us what to expect if we built the DC power supply in the lab and measured voltages using the oscilloscope and the Multimeter. These simulations give us practice in how to connect the oscilloscope and Multimeter instruments, and how to adjust these instruments in the lab so that we can measure the desired waveforms and voltages. We must connect and set up the virtual instruments the same way we would the real instruments in the lab. Thus, the virtual simulation serves several purposes: (1) It gives us practice in connecting a circuit using the same techniques we would in the lab, (2) it gives us practice connecting and adjusting instruments using the same techniques we would use in the lab, and (3) we get useful simulation results that can tell us how the circuit works. The simulation results can tell us if the circuit works properly or if our circuit operates within our design specifications.

The Transient Analysis simulation is used only to simulate the circuit, and the results are plotted on a graph rather than displayed on an instrument screen. The purpose of the Transient Analysis is to analyze the circuit and tell us if our design is working properly. Most SPICE simulators only have a Transient Analysis and do not have a virtual lab simulation using an oscilloscope. The results will be the same for both simulations. We are showing both the virtual simulation and the Transient Analysis to bridge the gap between the laboratory testing, the virtual lab simulations of Multisim, and traditional SPICE Transient Analysis simulations.

We will start with the circuit below, which is the circuit developed in Section 6.E:

Notice that the nodes at the input and output of the regulator have been labeled as **Vin** and **Vout**. To label a node, double-click the *LEFT* mouse button on a wire and then enter a name for the wire. Your labels may not be displayed in the same locations as shown in the screen capture above, but you do need to label the two nodes.

Next, we need to set up the Transient Analysis. Select **Simulate**, **Analyses**, and then **Transient Analysis** from the menus:

The **End time** is the total time of the simulation. The simulation runs from the **Start time** to the **End time**. We would like this simulation to run for 10 cycles of the 60 Hz waveform. The period of a 60 Hz waveform is 1/60 second. Ten periods is approximately 160 ms or 0.16 s.

The **Maximum time step settings** instruct Multisim on how to choose the time between simulation points. If you choose the option to **Generate time steps automatically**, Multisim will choose the time between simulation points to be as large as possible while keeping the simulation error below a specified maximum. This choice generates the least number of simulation points and allows the simulation to finish as quickly as possible. This is a good choice for long simulations that take several minutes or several hours to complete. The problem in using the **Generate time steps automatically** option is that simulation points are connected to one another using straight lines. With too few simulation points, curves tend to look jagged and you may question the accuracy of the simulation results since the actual waveforms are not jagged.

To avoid the problem of jagged curves resulting from too few simulation points, we can force the simulation to use more points using two options. The **Minimum number of time points** option allows us to specify a minimum number of points for the entire simulation. For example, if we specify this option to be 10,000, there will be at least 10,000 points during the simulation. Using this option instructs Multisim to use a maximum time between simulation points of

$$\text{maximum time between simulation points} = \frac{\text{TSTOP} - \text{TSTART}}{\text{minimum number of time points}}$$

For example, if we set **TSTART** to 0, **TSTOP** to 1 second, and specified a minimum of 1000 points, the largest time between simulation points would be 1 ms.

A second way to do the same thing is to specify a value for the **Maximum time step (TMAX)** option. This option directly specifies the maximum time between simulation points and is a standard parameter in most SPICE Transient Analyses. Thus, if we wanted the maximum time between simulation points to be 1 ms, we would specify this option to be 0.001.

For our first simulation of this DC power supply, we will not specify either of these options and will let Multisim choose the time step automatically. This will allow us to see how well Multisim chooses the time step and also see the effect of not specifying a maximum time between simulation points. Fill out the dialog box as shown below. Since each cycle takes about 16 ms, the simulation will run for about 10 cycles of the 60 Hz input voltage waveform.

Next, we must specify which traces we would like to look at. Select the **Output variables** tab:

Select to plot **vin** and **vout** during the simulation:

Click on the **Simulate** button to run the simulation. **vin** and **vout** will be displayed automatically by the Grapher:

The output voltage appears to be holding very constant at about 15 V as expected, but the input waveform does not appear to be quite right. The 120 Hz ripple is expected, but the ripple is not constant, and the envelope of the ripple seems to slowly oscillate. If we run the same simulation but change the end time to 0.4 seconds, we see that the ripple on the envelope does not appear to die out:

We see that the capacitor ripple never repeats itself, and this is not correct. From our understanding of the circuit, the observations in the lab, and the virtual simulation of the previous section, we know that the ripple at the input to the regulator should be periodic with a frequency of 120 Hz.

This simulation is incorrect because Multisim chooses too large a time between simulation points. We will fix this problem by specifying a maximum time step. Close the Grapher window by clicking on the [X] as shown below:

Select **Simulate**, **Analyses**, and then **Transient Analysis** from the menus to obtain the Transient Analysis setup dialog box:

We would like to specify a maximum time between simulation points, but how do we choose a value? A general method is to choose a large value for the maximum time and then run the simulation. Observe the results, and then rerun the simulation for a smaller value of the maximum time between points (a factor of 10 smaller is a good choice). At first you will notice a difference between the simulation results. As you make the maximum time step smaller, the results will eventually converge and you will notice no difference between simulation results even if you continue to reduce the maximum time between simulation points. When you reach this point, you have found a range of good values for the maximum time step, and you should use the largest value of the maximum time step such that the simulation results are the same. You can use a smaller value if you wish, but this could increase the amount of round-off error, take more simulation time, and enlarge the files that contain the calculated data.

I will not use this method here. Since I have a lot of experience simulating DC power supplies, I have a good idea of the value I would like to use (experience with running simulations also helps you choose the correct parameters for a simulation). I would like to have at least 100 simulation points in every 120 Hz cycle. The period of a 120 Hz waveform is 1/120 second, or 8.33 ms. To have 100 points occur in 8.33 ms, the maximum time between simulation points should be 8.33 ms divided by 100, or 0.083 ms, or 83e-6. Modify the dialog box as shown:

The dialog box is set up to simulate the circuit for approximately 10 cycles of a 60 Hz waveform, and the maximum time between simulation points is 0.083 ms, or about 100 points in a 120 Hz waveform.

The output variables should be specified from our previous simulation. However, we will check to make sure. Click on the **Output variables** tab and check to see if **vin** and **vout** are still specified to be plotted:

Click on the **Simulate** button to run the simulation. **vin** and **vout** will be plotted automatically:

The simulation looks better, but there does appear to be a slight variation in the envelope. I will run the same simulation but change the end time to 0.4 seconds.

The envelope of **vin** looks fairly constant, but to be careful, I will change the maximum time step to 10 μs and rerun the simulation. The results are shown below:

These results look about the same as when we had a maximum time step of 0.083 ms, so we conclude that the time step of 0.083 ms is sufficient. Closer inspection of the results may yield a difference, but for the plots shown, the results for a maximum time step of 10 μs and 0.083 ms appear to be about the same. Thus, we will rerun the simulation for a maximum time step of 0.083 ms and then examine the waveforms in a bit more detail. You can use the data generated with a time step of 10 μs, but your computer may take longer to plot the graphs because more data is being plotted. Note that the traces are thicker in the graph above because I changed the pen size in the Grapher.

We will now examine the ripple more closely. We would like to display the ripple on the input and the output and see how they relate to one another. To do this, we will create a new plot with the Postprocessor and plot the input and the output on separate graphs. If you were following the previous example, close the Grapher window by typing **CTRL-G**. We will start with Multisim displaying the circuit. We will assume that you have already run the simulation and that the data is available to the Postprocessor:

Select **Simulate** and then **Postprocess** from the menus:

Click on the **New Page** button and enter a name for the page:

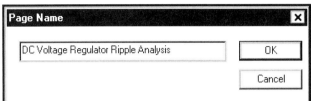

Click on the **OK** button to create the new page:

Next, we need to create a graph on this page. Click on the **New Graph** button and specify a name for the graph:

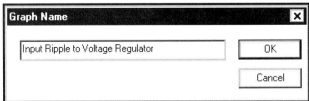

Click on the **OK** button to create the graph:

We now need to add the trace **vin** to this graph. Double-click on the text **vin** to select it and copy it to the trace:

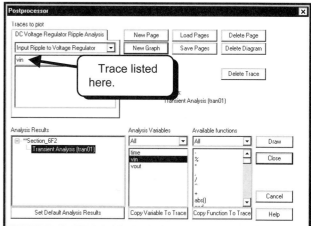

Click on the **Add Trace** button to add the trace to the graph.

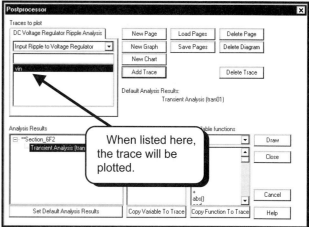

We now wish to create a second graph on the same page and add the output trace to the second graph. Click on the **New Graph** button and enter a name for the graph:

EWB Multisim — Time Domain Analyses 347

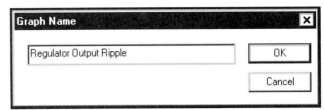

Click on the **OK** button to create the graph:

We now need to add the trace **vout** to this graph. Double-click on the text **vout** to select it and copy it to the trace:

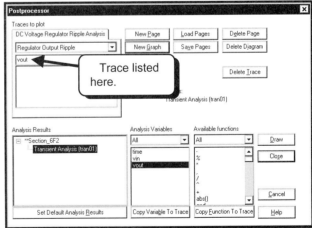

Click on the **Add Trace** button to add the trace to the graph.

We have now completed the setup for our plots. Click on the **Draw** button to view the page:

Close the **Postprocessor** window and enlarge the **Analysis Graphs** window:

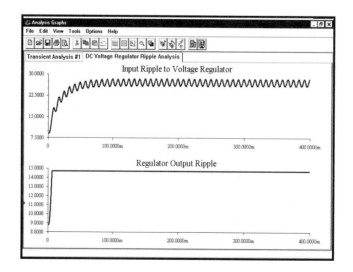

To see the ripple on the output we need to change the y-axis **Scale** of the bottom graph. To do this, click the **_RIGHT_** mouse button on the lower plot:

Select **Properties**:

This dialog box allows us to change all of the properties of the lower graph. For this example, all we need to do is change the y-axis scale. To do this, select the **Left Axis** tab:

From our virtual lab analysis of this circuit, we know that the average value of the output was 14.732 (see page 336) and peak-to-peak ripple was less than 5 mV (see page 334). From these numbers, we get a good idea of the y-axis range we need to see the ripple. Without this information, it may take a few guesses at the y-axis range before we find a suitable value. Specify the y-axis range to have a minimum of 14.6 and a maximum of 14.8 volts as shown below:

Click on the **OK** button to view the graph:

You can zoom in further if you wish. You can also use this technique to scale the time axis. To change the time axis, click the **RIGHT** mouse button on the graph and then select the **Bottom Axis** tab instead of the **Left Axis** tab. In the plot below, I have scaled the time axis of the top and the button graphs. The y-axis scales for the two plots are different, but the time axes for both plots have the same scale.

To get numerical values from the graph, you can use the cursors. To display the cursors, click the **RIGHT** mouse button on the bottom graph and then select **Toggle Cursors**. Use the mouse to place the cursors at the peaks of the waveforms.

From the **dy** value of the cursors, we see that the ripple at the regulator output is **4.8209** mV. To remove the cursors from the bottom graph, click the **RIGHT** mouse button on the bottom graph and then select **Toggle Cursors** again. Next, click the **RIGHT** mouse button on the top graph and then select **Toggle Cursors** to display the cursors on the top graph:

The **dy** value of the cursors shows us that the ripple at the regulator input is **2.4866** V.

Using these values to calculate the regulator ripple rejection ratio, we get

$$\text{Ripple Rejection Ratio}_{dB} = 20\log_{10}\left(\frac{\Delta Vin}{\Delta Vo}\right)$$

$$= 20\log_{10}\left(\frac{2.4866\,\text{V}}{4.8209\,\text{mV}}\right)$$

$$= 54.25\,\text{dB}$$

This value compares favorably to the value of 54 dB obtained using the virtual simulation on page 335.

EXERCISE 6-7: Run a power supply using a half-wave rectifier and a regulator. Use the same circuit as on page 327, except with a half-wave rectifier instead of a full-wave rectifier. Use the virtual diode because Multisim does not have a discrete diode that can handle the required current in the library. The remainder of the circuit should be the same.

SOLUTION:

Set up a Transient Analysis. Set the maximum time step to 10 μs to get accurate results. Also, set the end time to simulate several cycles to allow the capacitor to charge to a large enough voltage so that the regulator operates correctly:

The simulation results are:

We see that after several cycles, the input to the regulator becomes large enough so that the regulator functions properly and the regulator output is constant. We also note that after 0.6 seconds, the capacitor voltage (Vin) is still slowly increasing, but there is no need to run the simulation further because the output of the regulator is constant and the circuit appears to work properly.

We will zoom in to the beginning, middle, and end of the simulation to see some of the waveforms in more detail. The first 100 ms is shown below:

These waveforms show that the input to the regulator is too small, and the regulator cannot hold the output voltage constant. Another view is:

This view shows the input to the regulator becoming large enough so that the regulator can now function and keep the output voltage constant at 15 volts. An enlarged view of the end of the simulation is:

After the initial transient, the capacitor voltage is always large enough and the regulator functions properly so that the output voltage is constant at 15 V. The ripple voltage across the capacitor is about twice the magnitude of the ripple voltage of the full-wave rectifier. This is the expected result.

6.F. Zener Clipping Circuit – SPICE Transient Analysis

In this section we will demonstrate the use of a Zener diode, as well as the piece-wise-linear (PWL) voltage source. The PWL source can be used to create an arbitrary voltage waveform that is connected by straight lines between voltage points. Wire the circuit shown:

This circuit contains two new parts, the PWL voltage source and the Zener, so we will show how to place them in our circuit.

First we will place the PWL voltage source. Click the **LEFT** mouse button on the **Sources** button:

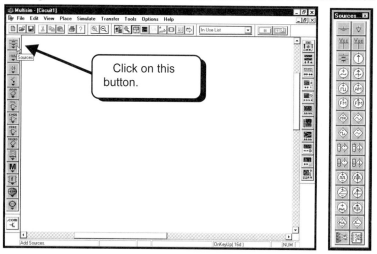

Click the **LEFT** mouse button on the **PIECEWISE_LINEAR_VOLTAGE_SOURCE** button. The source will become attached to the mouse:

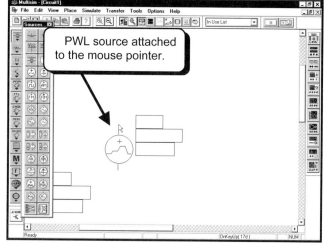

You can now place the PWL source in your circuit.

To place the Zener diode, click the **LEFT** mouse button on the **Diodes** button :

Click the **LEFT** mouse button on the **ZENER** diode button:

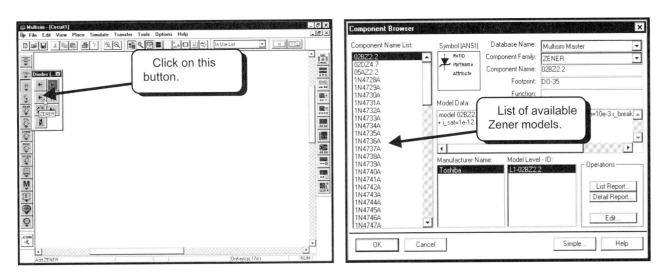

For this example we are using a **1N4734A** Zener diode, so select the diode and then click on the **OK** button. The diode will become attached to the mouse:

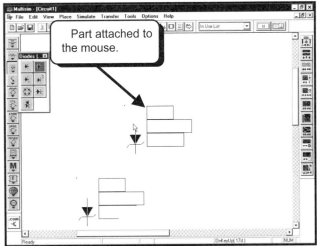

You can now place the Zener diode in your circuit. Continue drawing the circuit until you have the circuit shown below:

We now need to define the voltage waveform of the PWL voltage source. With a PWL voltage source, the time and voltage of points on the waveform are defined and then the points are connected by straight lines. This type of source is useful for creating waveforms made up of straight lines such as a saw tooth, ramp, or triangle wave. It is also useful if you have measured data of a waveform saved as time and voltage coordinates. The PWL source can be programmed with this data and the source will then reproduce the waveform. For this example, we will use the PWL source to create one cycle of a triangle wave.

To define the PWL waveform, double-click the **LEFT** mouse button on the PWL graphic:

We can define the source by either reading in a text file of data points or entering the data points manually using this dialog box. If you want to define the PWL source using a data file, the file should be a text file where each line contains a single point. The values on a single line must be separated by a space or a tab. The first value specified is the time, the second value is the voltage. An example is shown below:

```
0        0
0.001    15
0.003   -15
0.004    0
0.005    15
0.007   -15
0.008    0
0.009    15
0.011   -15
0.012    0
```

In the example above, the values are separated by a tab character. This data was created using Microsoft Excel and saved as a text file delimited by tabs. This is one of the standard formats available with Excel. You could also create this data file using the Windows Notepad and typing the data in manually. The data defines the following voltage waveform: At time = 0, the voltage is zero; at time = 0.001 s, the voltage is 15 volts; at time = 0.003 s, the voltage is –15 volts, and at time = 0.004, the voltage is zero volts. The waveform then repeats itself.

It appears that the PWL source does not have an easy way to generate a repeating waveform. To create a long repeating waveform using the PWL source, you would need to use Excel to create a large table of points and save the file as a text file with the tab delimiter. In Excel, you can easily create the points for a single period, and then use spreadsheet functions to make the data repeat periodically. We will not discuss this here.

For our example, we will simulate the circuit for a single period of a triangle wave. Thus, we will use the PWL source to generate a single triangle wave. We will enter the data points manually using the dialog box for the PWL source. The first time point (time = 0, voltage = 0) is already defined for us. To enter the second point, click the **LEFT** mouse button as shown below:

An empty cell appears. Enter the time coordinate for the second point. Do not press the **ENTER** key:

We now need to enter the voltage coordinate. Click the **LEFT** mouse button as shown below:

Type a value for the voltage coordinate of the point:

Repeat the process to add two more points: At time = 0.003 s, the voltage is −15 volts, and at time = 0.004 s, the voltage is 0 volts. The completed dialog box is shown below:

The data we entered defines the triangle wave below:

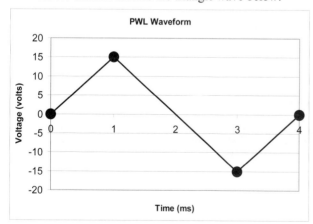

Table 6-1			
Time	Time	Time	Time
0	0.001	0.003	0.004
Voltage	Voltage	Voltage	Voltage
0	15	−15	0

Click on the **OK** button to accept the settings for the PWL source.

We would like to run a Transient Analysis for 4 ms. To set up the analysis select **Simulate**, **Analyses**, and then **Transient Analysis** from the Multisim menus:

We would like to run the Transient Analysis for 4 ms and to have at least 1000 simulation points during the simulation. Fill in the dialog box as shown:

Next, we must specify the waveforms we would like to see. Select the **Output variables** tab and specify that **vin** and **vout** be plotted during the simulation.

Click on the **Simulate** button to run the simulation:

I have turned on the grid by clicking the **RIGHT** mouse button on the graph and then selecting **Toggle Grid**. If you turn on the cursors (right mouse button and then **Toggle Cursors**) we can measure the maximum and minimum values of the output waveform:

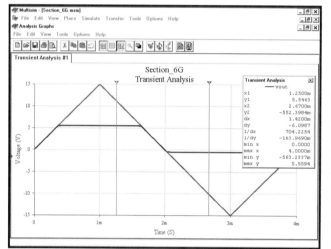

Cursor 1 (red) shows that the output is limited to a positive maximum value of about **5.5** volts and cursor 2 (blue) shows that the output is limited to a maximum negative voltage of **-552 m**V.

EXERCISE 6-8: Plot the input and output voltage waveforms using the Transient Analysis for the circuit below. Let the input be a ±15 volt triangle wave.

SOLUTION:

EXERCISE 6-9: Plot the input and output voltage waveforms using the Transient Analysis for the "Dead-Zone" clipping circuit below. Let the input be a ±15 volt triangle wave.

SOLUTION:

6.G. Zener Clipping Circuit – Virtual Lab Simulation

We will run the same simulation as in section 6.F, but we will use a square-wave voltage source to generate a triangle wave and use the oscilloscope to view the waveforms. Wire the circuit shown:

This circuit contains the pulse voltage source, which we have not yet used; so we will show how to place it in our circuit and set up its parameters. Click the **LEFT** mouse button on the **Sources** button:

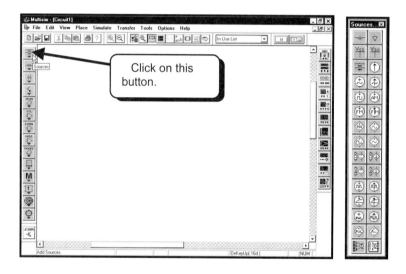

Click the **LEFT** mouse button on the **PULSE_VOLTAGE_SOURCE** button. The source will become attached to the mouse:

You can now place the source in your circuit. Continue drawing until you have the circuit shown below:

We now need to set the properties of the pulse voltage source. Double-click the **LEFT** mouse button on the pulse source graphic:

The **Initial Value** is the voltage level before the pulse starts, and it is also the voltage level after the pulse is finished. The **Pulsed Value** is the value of the voltage during the pulse. The **Pulse Width** is the amount of time the voltage stays at the **Pulsed Value**. The **Period** is the period of the waveform and is the amount of time over which the waveform repeats itself. The **Rise Time** is the amount of time it takes the voltage to go from the **Initial Value** to the **Pulsed Value**. The **Fall Time** is the amount of time it takes the voltage to go from the **Pulsed Value** to the **Initial Value**.

At the beginning of a simulation, the voltage starts at the **Initial Value**. The voltage then immediately starts rising towards the **Pulsed Value**. After the **Rise Time**, the voltage holds constant at the **Pulsed Voltage** for an amount of time equal to the **Pulse Width**. We see that the pulse starts immediately at the beginning of the simulation. If you want to delay the pulse at the beginning of a simulation, use the **Delay Time**. The **Delay Time** is the amount of time between the beginning of the simulation and the beginning of the first pulse. Its default value is zero so that the first pulse starts at the beginning of the simulation.

We would like to define a triangle wave that goes from −15 volts to 15 volts and has a **Period** of 4 ms. This triangle wave spends 2 ms going from −15 V to +15 V, thus we will specify the **Rise Time** as 2 ms. This triangle wave spends 2 ms going from +15 V to −15 V, so we will specify the **Fall Time** at 2 ms also. We will set the **Pulse Width** to zero because once the voltage reaches 15 V, it immediately reverses direction and heads back toward −15 V. If we specify the **Pulse Width** as something other than 0, when the voltage reaches 15 V, it will remain at 15 V for a length of time equal to the **Pulse Width**. This creates a flat line at +15 V that we do not want. The dialog box below defines a 15-volt amplitude triangle wave with a period of 4 ms:

Click on the **OK** button to accept the changes.

We will run a virtual simulation and view the results with the oscilloscope. Add the oscilloscope to your circuit as shown:

Run the simulation and then double-click on the oscilloscope to view the results:

The cursors show that the maximum positive value of vout is **5.6 V** and the maximum negative value is **-563.2 mV**. These values agree with the values obtained using the Transient Analysis on page 360.

6.G.1. Plotting Transfer Curves

When you have a clipping circuit, as in the last example, a transfer curve may be desired. A transfer curve is a plot of the output versus the input of a circuit. In the lab, this can be easily done using the X-Y capabilities of an oscilloscope. In normal mode, an oscilloscope displays **Channel A** versus time and **Channel B** versus time. When you switch the oscilloscope to X-Y mode, the oscilloscope will display **Channel A** versus **Channel B** or, depending on your scope, **Channel B** versus **Channel A**. The Multisim oscilloscope has the ability to display either **Channel A** versus **Channel B** or **Channel B** versus **Channel A**.

With a transfer curve, you usually plot the output voltage versus the input voltage. In our circuit, we are measuring **vin** with **Channel A** and **vout** with **Channel B**. We would like to plot **vout** on the y-axis and **vin** on the x-axis. This plot can be viewed on the oscilloscope by selecting the **B/A** button:

After selecting this button, you will see the transfer curve:

You can change the volts per division settings to enlarge the transfer curve. The **Channel A** settings control the x-axis and the **Channel B** settings control the y-axis. I will also use the cursors to measure the breakpoints of the transfer curve:

To return to the normal display of waveforms versus time, select the **Y/T** button. Note that selecting the **A/B** button will plot **Channel A** versus **Channel B**. For our connections, this would plot **Channel A** or **vin** on the y-axis and **Channel B** or **vout** on the x-axis. This would be a plot of **vin** versus **vout**, not quite the transfer curve we were looking for.

EXERCISE 6-10: Find the output voltage waveform and the transfer curve for the circuit below using the oscilloscope. Let the input be a ±15 volt triangle wave at 100 Hz. Use the pulsed voltage source to create the triangle wave.

SOLUTION: The attributes of the pulsed voltage source are:

Add the oscilloscope as shown:

The waveforms and transfer curve are:

EXERCISE 6-11: For the "Dead-Zone" clipping circuit shown below, find the output voltage waveform and the transfer curve. Let the input be a ±15 volt triangle wave at 500 Hz. Use the pulsed voltage source to create the triangle wave.

SOLUTION: The attributes of the pulsed voltage source are:

Add the oscilloscope as shown:

The waveforms and transfer curve are:

6.H. Amplifier Voltage Swing

In this section we will examine the voltage swing of the transistor amplifier used in section 5.C. In section 5.C we found that the gain of the amplifier was 45.7 dB, which corresponds to a gain of 193 V/V. We also found that the mid-band for this amplifier was from 65 Hz to 6.5 MHz. We would like to see what the maximum voltage swing of this amplifier looks like at mid-band. We will use the same circuit as in section 5.C.

Wire the following circuit:

The source is a sinusoidal voltage of amplitude 45 mV and frequency 1 kHz, $V_4 = 0.045\sin(2\pi*1000t)$. The amplitude was chosen to get the maximum swing out of this amplifier. You may have to have to run the simulation several times with different amplitude sine waves for **Vin** to see what amplitude you need for **Vin** to observe the maximum swing on the output. Too small an amplitude will cause a small output swing. Too large an output will cause the amplifier to go into saturation or cutoff. The amplitude should be chosen to cause the output of the amplifier to just saturate or cutoff.

We must now set up the parameters for the Transient Analysis. Select **Simulate**, **Analyses**, and then **Transient Analysis** from the Multisim menus. Set up the Transient Analysis simulation dialog box as shown below:

Note that we are running the simulation for 50 ms (**End time** = **0.05**). Since the source is at a frequency of 1 kHz, we are simulating the amplifier for 50 cycles. This is done to allow the amplifier to reach steady-state. If we look at the amplifier output at the beginning of the simulation, we will see the output voltage waveform slowly drift because the capacitors are charging in response to the DC average of the output voltage waveform. Since we are interested in the steady-state response of the amplifier, the simulation is run for 50 ms and we will look at the results near the end of the simulation. We have also specified the **Maximum time step** to be 5 µs to produce accurate simulation results. If you omit this value, Multisim may take large time steps and the waveform will look very jagged. A 1 kHz sine wave has a period of 1 ms or 1000 µs. With a **Maximum time step** of 5 µs, there will be at least 200 simulation points for each period of the sine wave. Specifying this small time step will result in smooth waveforms, but will also cause the simulation to take a long time to run.

The last thing we must do is select the **Output variables** tab of this dialog box and select **Vin** and **Vout** to be plotted during the simulation:

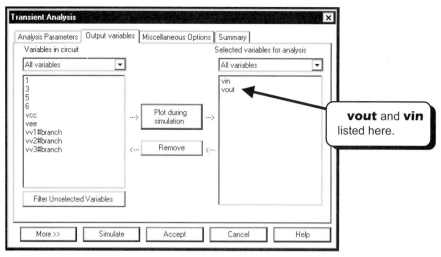

Click on the **Simulate** button to run the simulation:

The slow drift in the output waveform is easily seen in the screen capture above. The negative peak appears to slowly drift up during the first 10 ms, and then appears constant for the remainder of the simulation. To see the waveforms more closely, we will need to zoom in on the waveforms. To zoom in, **click and hold** the *LEFT* mouse button at the upper left corner of the area in which you want to zoom:

While continuing to hold down the mouse button, move the mouse pointer down and to the right. A zoom box will appear:

Move the mouse so that the zoom box encloses the portion of the waveform you want to view:

When you release the mouse button, Multisim will zoom in around the box:

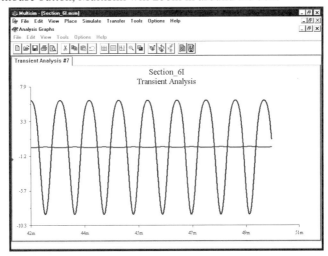

The output is so much larger than the input that you can hardly see any variation on the input. We also notice that the output does look sinusoidal, but also seems to be rounded at the top and pointed at the bottom. The output is not really a pure sine wave and contains some distortion. Distortion means that the output of the amplifier contains frequencies not present in the input signal. For a distortion-free amplifier, if we input a frequency of 1 kHz, the only frequency contained in the output will be 1 kHz. The output of our amplifier does not look like the input, thus it contains additional frequencies not present in the input. We will analyze the distortion in the output waveform in the next section using the Fourier Analysis.

If you want to see an enlarged view of the input voltage, you will need to plot the input and output voltages on separate graphs. See pages 344-350 to create two graphs and zoom in on those graphs. The input and output waveforms for the last 5 cycles of the simulation are shown below:

6.H.1. Fourier Analysis with Multisim

From the last simulation, we know that the amplifier adds distortion because the input to the amplifier was a pure sine wave at 1 kHz, and the output was a periodic wave, but it was not a pure sinusoid. From Fourier series analysis of periodic signals, we know that any periodic waveform is made up of a sum of sine waves at different frequencies and different phase angles. If our amplifier did not add distortion, the output would contain the same frequencies as the input. If we input a single frequency of 1 kHz, the output should only contain a single frequency at 1 kHz. Any distortion added by the amplifier would appear as additional frequencies in the output signal. The Fourier Analysis will tell us the frequency components of a signal, and thus can tell us the distortion added by the amplifier.

We will continue with the circuit from the previous section. To set up a Fourier Analysis, select **Simulate**, **Analyses**, and then **Fourier Analysis** from the menus:

Our input signal is a 1 kHz sine wave. This will be the lowest frequency in the output, and is called the fundamental frequency. Thus the default value of **1000 Hz** for the **Fundamental frequency** is correct. If our input signal was a different value, say 1.5 kHz, we would need to specify the **Fundamental frequency** as 1.5 kHz. The output waveform will contain multiples of the fundamental called harmonics. You can specify how many harmonics you would like to view. We will use the default value of 9; however, you can specify more or less. Your parameters should be set as shown above.

In order to calculate the Fourier components, Multisim must run a Transient analysis. Click on the **Edit transient analysis** button to set up the Transient Analysis:

We will set up the parameters for the Transient Analysis the same as in the previous section. The **End time** is set to 50 ms and the **Maximum time step** is set to 10 μs:

Click on the **Accept** button to return to the Fourier Analysis setup. Select the **Output variables** tab and then specify to plot **vout** during the simulation:

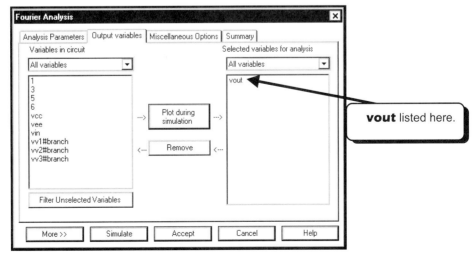

We are now ready to run the simulation, so click on the **Simulate** button:

The results are displayed as text and as a graph. The graph shows us that the output is composed mainly of two frequencies, the fundamental at 1 kHz and the second harmonic at 2 kHz:

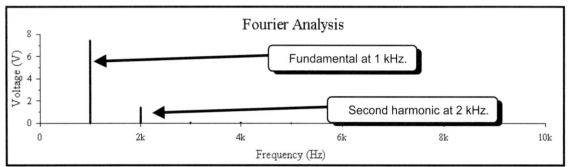

Other frequencies are contained in the output, but they are too small to see on the graph.

The text information gives us more detailed information. First, the total harmonic distortion (THD) is calculated as 18.6773%.

If we scroll down the text window, we get detailed information about each of the frequency components:

```
    No. Harmonics: 9, THD: 18.6773 %, Gridsize: 256, Interpolation Degree: 1

 Harmonic Frequency    Magnitude     Phase       Norm. Mag     Norm. Phase
 -------- ---------    ---------     -----       ---------     -----------
    1       1000        7.46686     -176.69          1              0
    2       2000        1.39336      97.9434      0.186605      274.635
    3       3000        0.0318442   166.108       0.00426474    342.799
```

4	4000	0.0497699	104.31	0.00666544	281.002
5	5000	0.00261411	44.1245	0.000350094	220.816
6	6000	0.00187943	108.075	0.000251703	284.767
7	7000	0.000218378	44.3525	2.92463e-005	221.044
8	8000	7.68373e-005	112.643	1.02904e-005	289.334
9	9000	1.95675e-005	86.4115	2.62057e-006	263.103

The results of the Fourier Analysis show that the magnitude of the sine wave at 1 kHz is **7.46686** V, and the magnitude of the sine wave at 2 kHz is **1.39336**. The text data also gives us the magnitude of the frequency components too small to see on the graph, and the phase of each frequency component. From the data above, an equation for the output voltage is:

$$V_o = 7.467 \sin(2\pi \bullet 1000t - 176.69°) + 1.393 \sin(2\pi \bullet 2000t + 97.94°) + 0.0318 \sin(2\pi \bullet 3000t + 166.1°) + \ldots$$

EXERCISE 6-12: Find the magnitude of the harmonics of a ±1 volt, 1 kHz square wave using the Fourier analysis.

SOLUTION: Use the pulse voltage source. Set the rise and fall times to 1 ns so that they are much shorter than the pulse width and period of the square wave. Wire the circuit below. The attributes for the pulse voltage source are also shown:

Set up the Fourier Analysis and Transient Analysis as shown:

The results are:

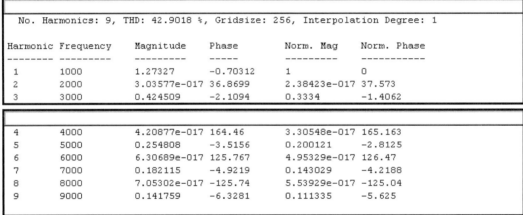

EXERCISE 6-13: Find the harmonics contained in the cross-over distortion of a push-pull amplifier. Let the input to the amplifier be a 2 V amplitude, 1 kHz sine wave. The push-pull amplifier drives a 100 Ω load. Run a Transient Analysis to view the waveforms and then run a Fourier Analysis to find the distortion.

SOLUTION: Wire the push-pull amplifier below. Run a Transient Analysis for one cycle with many points per cycle:

The input and output waveforms are:

EWB Multisim — Time Domain Analyses

Use the following setup for the Fourier Analysis. For the Transient Analysis within the Fourier Analysis, simulate the circuit for four cycles to provide more data for the Fourier series calculations.

The results of the Fourier Analysis are:

4	4000	0.00453029	-89.864	0.00417181	-89.864
5	5000	0.0880167	179.992	0.081052	179.992
6	6000	0.00249281	-89.669	0.00229555	-89.669
7	7000	0.0283589	179.98	0.0261148	179.98
8	8000	0.000715386	-89.018	0.000658778	-89.018
9	9000	0.00140767	179.861	0.00129628	179.861

We see that major frequency peaks of the output are at 1 kHz, 3 kHz, 5 kHz, and 7 kHz. Since the output contains frequencies not contained in the input, the amplifier adds distortion. The total harmonic distortion (THD) was **22.8679%**.

EXERCISE 6-14: Using feedback, we can reduce the distortion of the push-pull amplifier. Repeat the previous exercise using the amplifier below. An ideal OPAMP is used because a real OPAMP cannot drive this push-pull amplifier.

SOLUTION: Run a Transient Analysis for one cycle with many points per cycle. The input and output waveforms are:

We see that the input and output waveforms look exactly the same. This is because when you enclose an amplifier within a feedback loop, the feedback reduces the amplifier's distortion.

Use the following setup for the Fourier Analysis. For the Transient Analysis within the Fourier Analysis, simulate the circuit for four cycles to provide more data for the Fourier series calculations.

The results of the Fourier Analysis are:

```
No. Harmonics: 9, THD: 6.39063e-005 %, Gridsize: 256, Interpolation Degree: 1

Harmonic  Frequency    Magnitude      Phase         Norm. Mag      Norm. Phase
--------  ---------    ---------      -----         ---------      -----------
1         1000         1.99934        -1.2836e-006  1              0
2         2000         2.96847e-008   -86.961       1.48472e-008   -86.961
3         3000         5.02687e-007   -97.925       2.51426e-007   -97.925

4         4000         2.98179e-008   -83.952       1.49139e-008   -83.952
5         5000         5.74197e-007   134.554       2.87193e-007   134.554
6         6000         2.98474e-008   -80.941       1.49286e-008   -80.941
7         7000         6.09857e-007   43.5299       3.05029e-007   43.5299
8         8000         2.98551e-008   -77.936       1.49325e-008   -77.936
9         9000         8.21374e-007   -141.88       4.10823e-007   -141.88
```

We see that most of the distortion of the push-pull amplifier has been removed. The total harmonic distortion is about 6.4×10^{-5}%. This amplifier has two problems. The first is that most OPAMPs cannot drive this push-pull amplifier when it is driving a large load. The second is that the transistors turn on and off. This adds a phase delay, which can cause the amplifier to oscillate. A push-pull amplifier that is biased such that both transistors are on works better. This example was chosen to show how feedback can reduce distortion.

6.I. Ideal Operational Amplifier Integrator

In this section we will demonstrate the use of an ideal operational amplifier (OPAMP) and the pulsed voltage waveform. Wire the circuit shown below. The virtual OPAMPs are located in the Analog toolbar and are ideal OPAMPs unless you change the model.

To see the model used by the OPAMP, double-click the **LEFT** mouse button on the OPAMP graphic:

Select the **Edit Model** button:

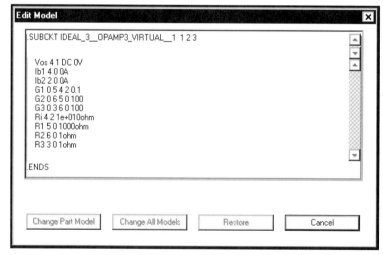

Vos is the offset voltage and is set to zero. **Ib1** and **Ib2** are the bias currents and they are also set to zero. **Ri** is the input impedance of the OPAMP, and its value is 1×10^{10} Ω. The gain of this subcircuit is made up of three cascaded voltage-controlled current sources, **G1**, **G2**, and **G3**. It is not obvious, but if you draw the circuit using the netlist shown, you will find that the gain is 1,000,000. These values make the OPAMP relatively ideal.

The drawback of using the ideal OPAMP model is that none of the non-ideal properties are modeled. In this example, if a non-ideal OPAMP model were used in the simulation, the integrator would not work because of bias currents. If this circuit were tested in the laboratory, it also would not work because of bias currents. Thus, the circuit simulation with a non-ideal OPAMP matches the results in the lab, but the circuit simulation with an ideal OPAMP does not match the lab results. For this example, the ideal model is not a good choice for simulation because it does not match the results in the lab.

We will use it here for demonstration purposes only. See **EXERCISE 6-16** to learn how this integrator performs using non-ideal OPAMPs. **In general, you should always use the non-ideal OPAMP models if possible**.

We will not change the model, so click on the **Cancel** button twice to return to the schematic.

The pulse voltage source can be used to create an arbitrary pulse-shaped waveform. We will use it to create a 1 kHz square wave with a 5 volt amplitude (the wave goes from –5 V to +5 V). The default settings for the pulse voltage source are a 0 to 5 volt square wave at 1 kHz. Thus, we must change the parameters. Double-click on the pulse voltage source graphic to modify its parameters. Set the parameters as shown below:

The parameters for the pulse voltage source were discussed on pages 362 to 364. The voltage waveform described by this dialog box will produce the following waveform: At the start of the simulation, the voltage will be set to the **Initial Value** of **5** volts. After the **Delay Time** (zero seconds in this case) the voltage will flip to the pulsed voltage of **–5** volts. From the **Delay Time** (t = 0) to 0.0005 s, the voltage will be equal to the **Pulsed Value** of **–5** volts. From 0.0005 s to t = 0.001 s, the voltage will be equal to the **Initial Value** of **5** V. At t = 0.001 s, the voltage will switch back to the **Pulsed Value** and repeat the cycle.

It is important to note that when Multisim calculates the bias to determine the initial condition of the capacitor, it sets the voltage of the pulsed source to its initial voltage. For our pulse waveform, Multisim will calculate the initial capacitor voltage assuming that Vin = 5 V. This will cause the integrator to have an infinite initial voltage. To prevent this, we must set the initial condition of the capacitor to 0 V. This can be done in several ways. We can set the initial condition of the capacitor to zero and then specify that Multisim use the specified initial conditions, or we can instruct Multisim to set all initial conditions to zero. When we set up the Transient Analysis, we will instruct Multisim to set all initial conditions to zero.

We would like to simulate the circuit for 10 cycles of the square wave using a Transient Analysis. Select **Simulate**, **Analyses**, and then **Transient Analysis** from the menus. Fill in the parameters as shown:

Note that we have specified that the initial conditions are set to zero. We will simulate the circuit for 10 cycles of the 1 kHz square wave or 10 ms. We have also specified a maximum step size of 10 μs to get an accurate simulation.

Select the **Output variables** tab and plot **vout** and **vin** during the simulation. Click on the **Simulate** button to run the Transient Analysis:

Note that the time integration of a square wave is a triangle wave.

EXERCISE 6-15: Run the integrator shown below. Set the initial condition of the capacitor to zero volts. Show the output waveform for a 1 kHz square wave input voltage. Note that this circuit will not work if the resistors are not matched exactly. This is a good circuit for simulation, but never use it in practice. You may also wish to run this circuit using non-ideal OPAMPs to see the effects of bias currents and offset voltages.

SOLUTION: The results of the Transient Analysis are:

EXERCISE 6-16: Run the integrator of the previous section using a non-ideal OPAMP model like the LM324J.

SOLUTION: Make sure you set the initial condition of the capacitor to zero. First, run the simulation for 10 ms:

The output voltage is:

At first glance, the graph looks the same as for the simulation for the ideal OPAMP. However, if you use the cursors to look at the positive peaks, you will see that they are slowly drifting positive:

At cursor 1, the peak value of the triangle wave is **2.8612** volts. At cursor 2, the peak of the triangle wave is **2.8792** volts, 0.018 volts higher than at the first cursor. Run the simulation again with the same settings as before, but set the end time to 100 ms and plot the output:

We see that the output is drifting quickly toward the positive supply. The screen captures below show a zoomed-in view of the drift at the start of the simulation (close to time = 0) and the drift at the end of the simulation (close to time = 100 ms).

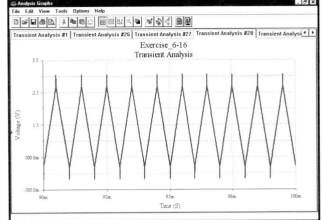

If we let the simulation run long enough, the OPAMP will eventually saturate and the output will be stuck at the positive supply rail. The output drifts up due to the bias currents of the OPAMP.

Note that the spikes at the top and the bottom of the output waveform should not be there. This may be an error in the OPAMP model:

6.J. Operational Amplifier Schmitt Trigger

In this section we will use an operational amplifier to create a Schmitt Trigger. A non-ideal operational amplifier must be used because the ideal OPAMP model has trouble converging when it is used as a Schmitt Trigger. Furthermore, a Schmitt Trigger requires an OPAMP with supply limits, so that the output of the circuit saturates at the supplies. Wire the circuit:

This OPAMP circuit is a Schmitt Trigger with trigger points at approximately ±7.5 V. A sinusoidal voltage source will be used to swing the input from +14 V to –14 V and from –14 V to +14 V a few times. The frequency of the source is 1 Hz to eliminate the effects of the OPAMP slew rate on the Schmitt Trigger performance. If you wish to observe slew rate effects, a higher frequency should be chosen. We would also like the sinusoidal source to start at –14 V instead of zero. This can be done by setting the phase of the source to 90°.

Since many of the attributes of the sinusoidal source have not been previously described, we will give a brief description here. The attributes of a sinusoidal source are:

[AC Voltage dialog box: Voltage Amplitude 14 V, Voltage RMS 9.90 V, Tolerance 33.3 %, Voltage Offset 0 V, Frequency 1 Hz, Time Delay 0 sec, Damping Factor 0 1/sec, Phase 90 Deg]

The sinusoidal source is best described by an equation. We will use the following variables:

Attribute	Symbol	Units
Phase	φ	degrees
Voltage Offset	B	volts
Voltage Amplitude	A	volts
Frequency	F	Hz
Damping Factor	β	$\sec^{-1}$
Delay Time	τ_d	seconds

Using these variables, the equation for the sinusoidal source is:

$$V_{\sin} = \begin{cases} B - A\sin\left(\left(\dfrac{2\pi}{360}\right)\varphi\right) & \text{for } 0 \le t < \tau_d \\ B + A\sin\left[2\pi\ F(t - \tau_d) - \left(\dfrac{2\pi}{360}\right)\varphi\right] e^{-\beta(t-\tau_d)} & \text{for } t \ge \tau_d \end{cases}$$

(This term is a constant.)

This may not be too clear. The full sinusoidal source can be exponentially damped, have a DC offset, and have a time delay as well as a phase delay. In the above equation, the phase (ϕ) is specified in degrees and is converted into radians by the constant 2π/360. We note that for $\tau_d \ge t \ge 0$, $V_{\sin}$ is constant. The sinusoid does not start until t = τ_d. If there are no time or phase delays, the above equation reduces to the exponentially damped sine wave:

$$V_{\sin} = B + A\sin[2\pi\ Ft]e^{-\beta t}$$

The sources available for use in the Transient Analysis are very flexible to provide the user with many possible waveforms for simulation. Unfortunately, this flexibility can also lead to confusion when using a source for the first time.

We would now like to set up a Transient Analysis that allows the sinusoidal source to complete two cycles. Since the source has a frequency of 1 Hz, two cycles will take two seconds of simulation time. Select **Simulate**, **Analyses**, and then **Transient Analysis** from the menus. Fill in the parameters as shown:

EWB Multisim — Time Domain Analyses

We have specified a **Maximum time step** of 1 ms to get accurate results when the Schmitt Trigger flips.

Next, select the **Output variables** tab and specify to plot **vin** and **vout**:

Click on the **Simulate** button to run the simulation:

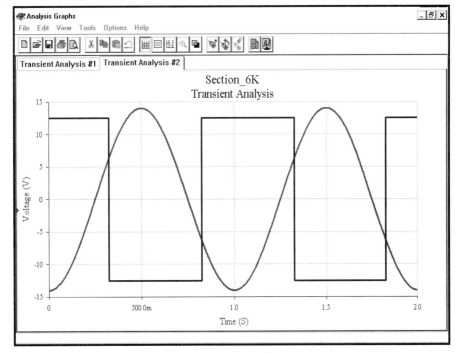

Using the cursors, we can display the trigger points of the Schmitt Trigger:

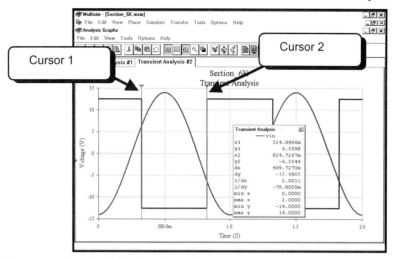

We can see that the Schmitt Trigger changes state at approximately VIN = ±6.3 V.

The next thing we would like to do is plot the hysteresis curve for this Schmitt Trigger. The hysteresis curve is a plot of V_O versus V_{IN}. To do this, we must display the waveforms on the Oscilloscope instrument and then change the mode of the scope. Add the oscilloscope to the circuit and then connect **Vin** to **Channel A** and **Vout** to **Channel B**. Change the frequency **V1** to **10 Hz**:

Click on the Run button to start the simulation and then double-click on the Oscilloscope instrument to view the waveforms. You will need to change the oscilloscope **Timebase** to 20 ms/div, and the oscilloscope window will take a long time to display the waveforms. After a while, you will see the waveforms below:

EWB Multisim   Time Domain Analyses   389

To view the hysteresis curve, all we need to do is to change the oscilloscope mode to **B/A**. Click on the **B/A** button as shown:

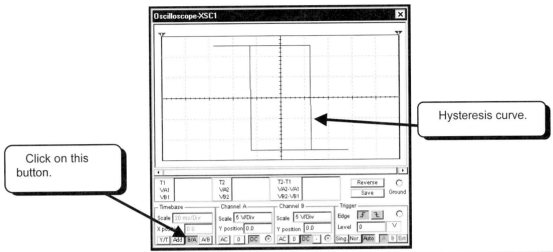

EXERCISE 6-17: Run a non-inverting Schmitt Trigger with a ±14 volt sine wave input.

SOLUTION: Use the sinusoidal voltage source. The input and output waveforms are shown on the left screen capture, and the hysteresis curve is shown on the right screen capture.

6.K. Parameter Sweep — Inverter Switching Speed

When designing digital circuits we are concerned with how different circuit elements affect the operation of the circuit. In this section we will look at switching speed. First, we will look at a basic BJT inverter and observe its operation. Wire the circuit below:

The input to the inverter will be a short 1 μs pulse. Double-click the **LEFT** mouse button on the pulse voltage source graphic to view the parameters for this source. Fill in the parameters as shown:

The **Pulse Width** of the pulse is 1 μs and the **Period** is 50 μs. These settings produce a short 1 μs pulse that repeats every 50 μs. We will only run the simulation for 50 μs, so we will only see a single pulse. Normally, the pulse will start at the beginning of the simulation. However, we have specified a **Delay Time** of 1 μs, so the voltage stays at the **Initial Value** of 0 V for 1 μs before the source starts the pulse. The **Rise Time** and **Fall Time** are set to 1 ns because we are looking at the switching speed of a logic gate and we want the edges of our input pulse to be much faster than the rise and fall times of the circuit we are testing.

We will use a Transient Analysis here instead of a virtual simulation because we are going to look at the response of this circuit for several different values of the collector resistor, RC. We cannot do this with a virtual simulation. We will first run a Transient Analysis to look at the response for a single value of RC. Select **Simulate**, **Analyses**, and then **Transient Analysis** from the menus and fill in the dialog box as shown:

Notice that we have set the **End time** to 50 μs and have specified the **Maximum time step** to be 10 ns.

We will first look at the input voltage pulse to see that it is correct. Select the **Output variables** tab and then specify to plot **vin** during the simulation:

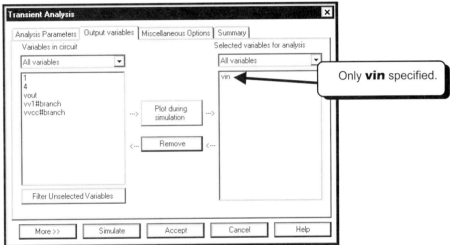

Click on the **Simulate** button to run the simulation. Your input voltage waveform should be as shown below:

We see a 0 to 5 V pulse that lasts for 1 μs. We also note that the pulse starts after a delay of 1 us, as specified by the Delay Time in the pulse voltage source parameters. If your pulse does not look as shown above, modify the source parameters and rerun the simulation.

Now that we have verified the input pulse waveform, we need to test the operation of the circuit and plot the output. Select **Simulate**, **Analyses**, and then **Transient Analysis** from the menus and verify that your dialog box still has the same settings as shown below:

Select the **Output variables** tab and change the settings so that only **vout** will be plotted during the simulation:

Click on the **Simulate** button to run the simulation and view the output voltage waveform:

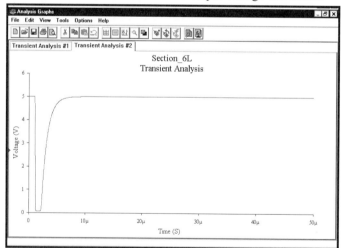

We see that the fall time of the output (the output goes from 5 V down to 0 V) is very fast, but the circuit has a slow rise time (the output goes from 0 V (or close to it) up to 5 V).

We would now like to answer the question of how the rise and fall times of the output of the circuit are affected by the value of the collector resistor. We will answer this question using the Parameter Sweep. Select **Simulate**, **Analyses**, and then **Parameter Sweep** from the menus and fill in the dialog box as shown:

The top portion of the dialog box specifies that we are going to change the resistance of resistor RC. The bottom portion of the dialog box specifies the values that RC will have. In this case we have specified a linear sweep from 1000 Ω to 10,000 Ω in 1000 Ω steps. Thus, RC will have values of 1000 Ω, 2000 Ω, 3000 Ω, and so on, up to 10,000 Ω.

The next thing we must do is specify that the Parameter Sweep works in conjunction with a Transient Analysis. Select the **More** button:

By default, the **Transient analysis** is selected. If not, select it.

We also note that the option **Group all traces on one plot** is selected. With this option selected, all runs from the Parameter Sweep will be plotted on a single graph. For our example, the resistor value takes on ten values. When we plot any trace, ten traces will be plotted, one for each value of the resistor. If we plot the input voltage, ten traces will be drawn on the graph. The ten traces will be the input voltages for all the different values of the resistor. If we plot the output voltage, ten traces will also be drawn. Each trace will be the output voltage for each value of the resistor.

The next thing we must do is edit the parameters for the Transient Analysis. Select the **Edit Analysis** button and fill in the parameters with the values we used in our analysis at the beginning of this section:

Click on the **Accept** button to return to the **Parameter Sweep** setup dialog box:

The last thing we must do is specify the **Output variables** for the run. Select the **Output variables** tab and specify that **vout** be displayed during the simulation:

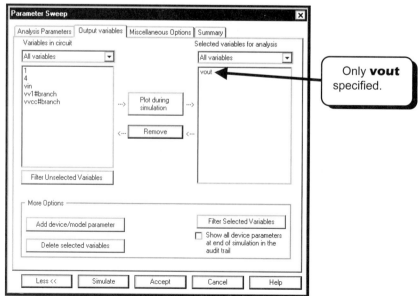

We are now ready to run the simulation, so click on the **Simulate** button:

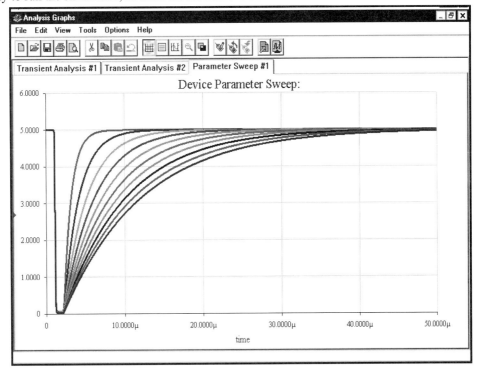

We see that the fall time is independent of our choice of RC. With the time scale shown, all of the runs go from 5 V down to 0 V so fast that it looks like a square edge. The rise time, however, appears to be a direct function of our choice of RC.

The question may arise as to how we tell which trace is for what value of RC. This can be done by displaying the legend. Click the **RIGHT** mouse button on the graph and then select **Toggle Legend** to display the legend:

With the legend displayed, we can easily identify each trace, and we see that a smaller value of RC yields a shorter rise time. Thus, to make a logic gate of this type with a fast rise time, we need to design our circuit using a small value of RC.

EXERCISE 6-18: For the BJT inverter of this section, run the circuit to observe how Rb affects switching speed. Let RC remain constant at 1 kΩ.

SOLUTION: Modify the Parameter Sweep to sweep resistor RB from 5 kΩ to 30 kΩ in 5 kΩ steps.

Modify the Transient Analysis setup to run the simulation for 7.5 μs and use a **Maximum step size** of 1 ns:

The results of the simulation are:

The results show that the base resistance has a direct effect on the fall time and a small effect on the rise time.

6.L. Temperature Sweep — Linear Regulator

A temperature sweep can be used in conjunction with most of the analyses, and it allows you to run the same analysis several times at different temperatures. Here we will show how temperature affects the performance of a linear voltage regulator. Create the circuit shown below:

Time Domain Analyses

Voltage source **V1** represents the output of a 12.6 VRMS transformer when it is connected to the 115 VRMS line. This circuit is a bridge rectifier followed by a filter capacitor to produce a DC voltage with ripple at **Vin**. After **Vin** is a linear regulator made from a Zener voltage reference and an NPN pass transistor. This regulator should produce a DC output voltage of approximately 5 volts. We will first run a Transient Analysis to see the operation of the circuit at room temperature (27°C). To set up a Transient Analysis, select **Simulate, Analyses**, and then **Transient Analysis** from the menus. Fill in the parameters as shown below:

Select the **Output variables** tab and specify to display **vin** and **vout** during the simulation:

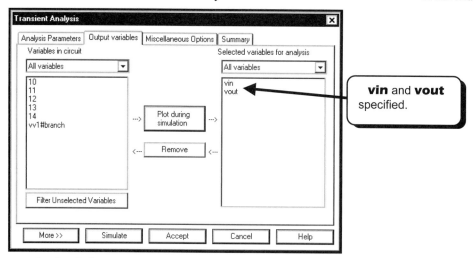

Select the **Simulate** button to run the simulation:

We will zoom in on the **vout** trace to see it more closely. Use the cursors to find the magnitude of the ripple.

We see that the output voltage is about 4.7 V and the magnitude of the ripple on the output is about 25 mV.

Next we will see how the output changes with temperature. We will run a Transient Analysis at −25°C, 25°C, and 125°C. Return to the schematic and select **Simulate**, **Analyses**, and then **Temperature Sweep** from the menus, and fill in the **Temperature Sweep Analysis** dialog box as shown:

EWB Multisim — Time Domain Analyses

We see that we have specified the **Sweep Variable Type** as a **List**. This option allows us to specify a list of temperatures at which we want to run the simulation. There does not need to be any relationship between the temperatures in the list.

The Temperature Sweep Analysis must run in conjunction with another analysis. To specify the analysis, click on the **More** button:

By default, a **Transient analysis** is selected. Had you wanted to run an AC Analysis or a DC Operating Point analysis, you would select that analysis here. We need to set up the parameters for the Transient Analysis, so select the **Edit Analysis** button and fill in the parameters with the values we used for the Transient Analysis at the beginning of this section:

Click on the **Accept** button to return to the **Temperature Sweep Analysis** dialog box. Select the **Output variables** tab and specify to plot **vin** and **vout** during the simulation:

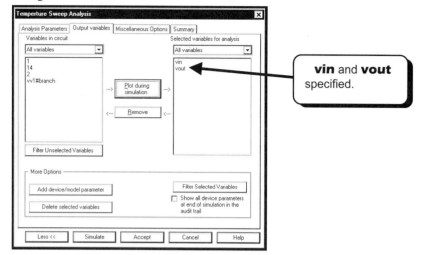

Click on the **Simulate** button to run the simulation. It will take three times as long as the previous simulation because the Transient Analysis will run three times. Logically, the Transient Analysis runs inside the Temperature Sweep. For this example, the temperature will first be set to –25°C, and then the Transient Analysis will be run. Next, the temperature will be set to 25°C, and then the Transient Analysis will be run. Finally, the temperature will be set to 125°C, and then the Transient Analysis will be run.

Three traces are shown for each variable, one trace for each simulation temperature. We see that both the input and the output voltage waveforms are affected by temperature. We will first look at the input voltage waveforms more closely:

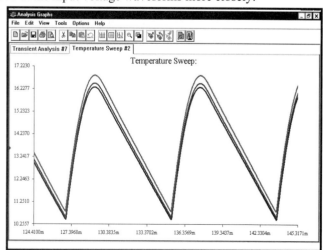

The variation in the input voltage waveform is mostly due to the temperature dependence of the diodes in the bridge rectifier. The voltage regulator also draws a slightly different current at different temperatures, and this also changes the input voltage waveform, but not nearly as much as bridge diodes.

Next, we will look at the output voltage traces. Three traces are shown, one for each temperature.

We can use the cursors to measure the ripple on each trace:

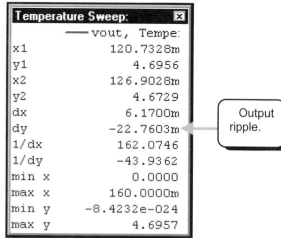

The top trace has 31 mV of ripple, the middle trace has 26 mV of ripple, and the bottom trace has 23 mV of ripple.

The question may arise as to how to tell which trace is at what temperature. To identify the traces, click the **RIGHT** mouse button on the graph and then select **Toggle Legend**:

The legend tells us which trace is **vin** and **vout**, and also tells us the temperature of each trace:

6.M. Problems

Problem 6-1: Find the phase in degrees of the voltage at node V1 using the Oscilloscope instrument in the circuit below:

Problem 6-2: For the circuit of Problem 6-1, find the phase in degrees of the voltage at node V2 and the current through R1 using the Oscilloscope instrument.

Problem 6-3: For the circuit of Problem 6-1, find the phase in degrees of the voltage at node V3 and the current through C2 using the Oscilloscope instrument.

Problem 6-4: For the circuit of Problem 6-1, find the phase in degrees of the voltage across R1 and the current through C2 using the Oscilloscope instrument.

Problem 6-5: For the circuit of Problem 6-1, find the phase in degrees of the voltage at node V1 and the current through R2 using the Oscilloscope instrument.

Problem 6-6: For the circuit of Problem 6-1, find the phase in degrees of the voltage at node V2 and the current through C1 using the Oscilloscope instrument.

Problem 6-7: For the circuit of Problem 6-1, find the phase in degrees of the voltage at node V3 and the current through R3 using the Oscilloscope instrument.

Problem 6-8: For the circuit of Problem 6-1, find the phase in degrees of the voltage across C2 and the current through C2 using the Oscilloscope instrument.

Problem 6-9: Find the phase in degrees of the voltage at node V1 using indicators in the circuit below:

Problem 6-10: For the circuit of Problem 6-9, find the phase in degrees of the voltage at node V2 and the current through R1 using the Oscilloscope instrument.

Problem 6-11: For the circuit of Problem 6-9, find the phase in degrees of the voltage at node V3 and the current through C2 using the Oscilloscope instrument.

Problem 6-12: For the circuit of Problem 6-9, find the phase in degrees of the voltage at node V1 and the current through R2 using the Oscilloscope instrument.

Problem 6-13: For the circuit of Problem 6-9, find the phase in degrees of the voltage at node V2 and the current through C1 using the Oscilloscope instrument.

Problem 6-14: For the circuit of Problem 6-9, find the phase in degrees of the voltage across C2 and the current through C1 using the Oscilloscope instrument.

Problem 6-15: For the circuit of Problem 6-9, find the phase in degrees of the voltage at node V3 and the current through L1 using the Oscilloscope instrument.

Problem 6-16: Find the phase in degrees of the voltage at node V1 using the Oscilloscope instrument in the circuit below:

Problem 6-17: For the circuit of Problem 6-16, find the phase in degrees of the voltage at node V2 and the current through R1 using the Oscilloscope instrument.

Problem 6-18: For the circuit of Problem 6-16, find the phase in degrees of the voltage at node V3 and the current through C2 using the Oscilloscope instrument.

Problem 6-19: For the circuit of Problem 6-16, find the phase in degrees of the voltage at node V1 and the current through R2 using the Oscilloscope instrument.

Problem 6-20: For the circuit of Problem 6-16, find the phase in degrees of the voltage across L1 and the current through R2 using the Oscilloscope instrument.

Problem 6-21: For the circuit of Problem 6-16, find the phase in degrees of the voltage at node V2 and the current through C1 using the Oscilloscope instrument.

Problem 6-22: For the circuit of Problem 6-16, find the phase in degrees of the voltage at node V3 and the current through L1 using the Oscilloscope instrument.

Problem 6-23: For the full-wave power supply shown on page 327, what happens to the voltages Vin and Vout if capacitor C1 is made larger?

Problem 6-24: For the full-wave power supply shown on page 327, what happens to the voltages Vin and Vout if capacitor C1 is made smaller? What is the smallest value of C1 such that the output voltage remains constant?

Problem 6-25: For the full-wave power supply shown on page 327, what happens to the diode currents if capacitor C1 is made larger?

Problem 6-26: For the full-wave power supply shown on page 327, what happens to the diode currents if capacitor C1 is made smaller?

Problem 6-27: For the full-wave power supply shown on page 327, what happens to the voltages Vin and Vout if the load is made smaller (higher resistance)?

Problem 6-28: For the full-wave power supply shown on page 327, what happens to the voltages Vin and Vout if the load is made larger (lower resistance)? What is the largest load such that the output voltage remains constant?

Problem 6-29: For the full-wave power supply shown on page 327, what happens to the diode currents if the load is made larger (lower resistance)?

Problem 6-30: For the full-wave power supply shown on page 327, what happens to the diode currents if the load is made smaller (higher resistance)?

Problem 6-31: For the half-wave power supply shown on page 337, what happens to the voltages Vin and Vout if capacitor C1 is made larger?

Problem 6-32: For the half-wave power supply shown on page 337, what happens to the voltages Vin and Vout if capacitor C1 is made smaller? What is the smallest value of C1 such that the output voltage remains constant?

Problem 6-33: For the half-wave power supply shown on page 337, what happens to the diode currents if capacitor C1 is made larger?

Problem 6-34: For the half-wave power supply shown on page 337, what happens to the diode currents if capacitor C1 is made smaller?

Problem 6-35: For the half-wave power supply shown on page 337, what happens to the voltages Vin and Vout if the load is made smaller (higher resistance)?

Problem 6-36: For the half-wave power supply shown on page 337, what happens to the voltages Vin and Vout if the load is made larger (lower resistance)? What is the largest load such that the output voltage remains constant?

Problem 6-37: For the half-wave power supply shown on page 337, what happens to the diode currents if the load is made larger (lower resistance)?

Problem 6-38: For the half-wave power supply shown on page 337, what happens to the diode currents if the load is made smaller (higher resistance)?

Problem 6-39: For the circuit below let Vin be a triangle wave that varies between –20 and +20 volts. (a) Plot Vo(t) and Vin(t) for one cycle of the input waveform using a Transient Analysis. (b) Plot the transfer curve Vo versus Vin using the Oscilloscope instrument.

Problem 6-40: For the circuit of Problem 6-39 let Vin be a triangle wave that varies between –20 and +20 volts. (a) Plot Vo(t) and Vin(t) for about one cycle of the input waveform using the Oscilloscope instrument. (b) Plot the transfer curve Vo versus Vin using the Oscilloscope instrument.

Problem 6-41: For the circuit below let Vin be a triangle wave that varies between –20 and +20 volts. (a) Plot Vo(t) and Vin(t) for one cycle of the input waveform using a Transient Analysis. (b) Plot the transfer curve Vo versus Vin using the Oscilloscope instrument.

Problem 6-42: For the circuit of Problem 6-41 let Vin be a triangle wave that varies between –20 and +20 volts. (a) Plot Vo(t) and Vin(t) for about one cycle of the input waveform using the Oscilloscope instrument. (b) Plot the transfer curve Vo versus Vin using the Oscilloscope instrument.

Problem 6-43: For the circuit below let Vin be a triangle wave that varies between –20 and +20 volts. (a) Plot Vo(t) and Vin(t) for one cycle of the input waveform using a Transient Analysis. (b) Plot the transfer curve Vo versus Vin using the Oscilloscope instrument.

Problem 6-44: For the circuit of Problem 6-43 let Vin be a triangle wave that varies between –20 and +20 volts. (a) Plot Vo(t) and Vin(t) for about one cycle of the input waveform using the Oscilloscope instrument. (b) Plot the transfer curve Vo versus Vin using the Oscilloscope instrument.

Problem 6-45: For the circuit below let Vin be a triangle wave that varies between –20 and +20 volts. (a) Plot Vo(t) and Vin(t) for one cycle of the input waveform using a Transient Analysis. (b) Plot the transfer curve Vo versus Vin using the Oscilloscope instrument.

Problem 6-46: For the circuit of Problem 6-45 let Vin be a triangle wave that varies between –20 and +20 volts. (a) Plot Vo(t) and Vin(t) for about one cycle of the input waveform using the Oscilloscope instrument. (b) Plot the transfer curve Vo versus Vin using the Oscilloscope instrument.

Problem 6-47: For the circuit below let Vin be a triangle wave that varies between –20 and +20 volts. (a) Plot Vo(t) and Vin(t) for one cycle of the input waveform using a Transient Analysis. (b) Plot the transfer curve Vo versus Vin using the Oscilloscope instrument.

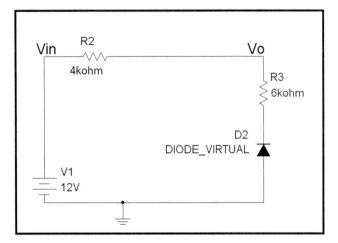

Problem 6-48: For the circuit of Problem 6-47 let Vin be a triangle wave that varies between –20 and +20 volts. (a) Plot Vo(t) and Vin(t) for about one cycle of the input waveform using the Oscilloscope instrument. (b) Plot the transfer curve Vo versus Vin using the Oscilloscope instrument.

Problem 6-49: For the circuit below let Vin be a triangle wave that varies between –20 and +20 volts. (a) Plot Vo(t) and Vin(t) for one cycle of the input waveform using a Transient Analysis. (b) Plot the transfer curve Vo versus Vin using the Oscilloscope instrument.

Problem 6-50: For the circuit of Problem 6-49 let Vin be a triangle wave that varies between –20 and +20 volts. (a) Plot Vo(t) and Vin(t) for about one cycle of the input waveform using the Oscilloscope instrument. (b) Plot the transfer curve Vo versus Vin using the Oscilloscope instrument.

Problem 6-51: For the circuit below let Vin be a sine wave at 1000 Hz. (a) Find the maximum voltage swing at Vout without clipping or a large amount of distortion. (b) What is the amplitude of Vin and Vout found in part a? (c) Run a Fourier Analysis with the output at its maximum found in part a. What is the total harmonic distortion contained in the output? What is the magnitude of the largest harmonic?

Problem 6-52: Repeat the simulations of Problem 6-51 with RL = 10 kΩ.

Problem 6-53: Repeat the simulations of Problem 6-51 with RL = 100 Ω.

Problem 6-54: Repeat the simulations of Problem 6-51 with RL = 10 Ω.

Problem 6-55: Repeat the simulations of Problem 6-51 with RC = 0 Ω. What is the effect of RC on this circuit?

Problem 6-56: For the circuit below let Vin be a sine wave at 1000 Hz. (a) Find the maximum voltage swing at Vout without clipping or a large amount of distortion. (b) What is the amplitude of Vin and Vout found in part a? (c) Run a Fourier Analysis with the output at its maximum found in part a. What is the total harmonic distortion contained in the output? What is the magnitude of the largest harmonic?

Problem 6-57: Repeat the simulations of Problem 6-56 with RL = 10 kΩ.

Problem 6-58: Repeat the simulations of Problem 6-56 with RL = 100 Ω.

Problem 6-59: Repeat the simulations of Problem 6-56 with RL = 10 Ω.

Problem 6-60: For the circuit below let Vin be a sine wave at 1000 Hz. (a) Find the maximum voltage swing at Vout without clipping or a large amount of distortion. (b) What is the amplitude of Vin and Vout found in part a? (c) Run a Fourier Analysis with the output at its maximum found in part a. What is the total harmonic distortion contained in the output? What is the magnitude of the largest harmonic?

Problem 6-61: Repeat the simulations of Problem 6-60 with RL = 1 kΩ.

Problem 6-62: Repeat the simulations of Problem 6-60 with RL = 100 Ω.

Problem 6-63: Repeat the simulations of Problem 6-60 with RL = 10 Ω.

Problem 6-64: For the circuit below let Vin be a sine wave at 1000 Hz. (a) Find the maximum voltage swing at Vout without clipping or a large amount of distortion. (b) What is the amplitude of Vin and Vout found in part a? (c) Run a Fourier Analysis with the output at its maximum found in part a. What is the total harmonic distortion contained in the output? What is the magnitude of the largest harmonic?

Problem 6-65: Repeat the simulations of Problem 6-64 with RL = 10 kΩ.

Problem 6-66: Repeat the simulations of Problem 6-64 with RL = 100 Ω.

Problem 6-67: Repeat the simulations of Problem 6-64 with RL = 10 Ω.

Problem 6-68: Find the transfer curve for the Schmitt Trigger shown below:

Problem 6-69: Find the transfer curve for the Schmitt Trigger shown below. The input voltage waveform should be limited to vary between 0 and 5 volts.

Problem 6-70: Find the transfer curve for the Schmitt Trigger shown below.

Problem 6-71: Find the transfer curve for the Schmitt Trigger shown below.

Problem 6-72: Find the transfer curve for the Schmitt Trigger shown below.

Problem 6-73: For the circuit below, run a Parameter Sweep in conjunction with a Transient Analysis to see how varying Rd from 1 kΩ to 10 kΩ in 1 kΩ steps affects the rise and fall times of the output waveform.

Problem 6-74: For the circuit of Problem 6-73:

(a) Run a Parameter Sweep in conjunction with a Transient Analysis to see how varying Rg from 10 Ω to 100 Ω in 10 Ω steps affects the rise and fall times of the output waveform.

(b) Run a Parameter Sweep in conjunction with a Transient Analysis to see how varying Rg from 10 Ω to 100 Ω in 10 Ω steps affects the gate-source voltage of the MOSFET.

Problem 6-75: For the circuit of Problem 6-73:

(a) Run a Parameter Sweep in conjunction with a Transient Analysis to see how varying Rg from 100 Ω to 500 Ω in 100 Ω steps affects the rise and fall times of the output waveform.

(b) Run a Parameter Sweep in conjunction with a Transient Analysis to see how varying Rg from 100 Ω to 500 Ω in 100 Ω steps affects the gate-source voltage of the MOSFET.

Problem 6-76: For the circuit below, run a Parameter Sweep in conjunction with a Transient Analysis to see how varying R from 1 kΩ to 10 kΩ in 1 kΩ steps affects the rise and fall times of the output waveform. V1 is a 0 to 15 volt square wave. You must choose the appropriate parameters for the pulse voltage source so that the output has time to fully charge and discharge.

Problem 6-77: For the circuit of Problem 6-76, run a Parameter Sweep in conjunction with a Transient Analysis to see how varying R from 100 Ω to 1 kΩ in 100 Ω steps affects the rise and fall times of the output waveform. V1 is a 0 to 15 volt square wave. You must choose the appropriate parameters for the pulse voltage source so that the output has time to fully charge and discharge.

Problem 6-78: For the circuit of Problem 6-76, run a Parameter Sweep in conjunction with a Transient Analysis to see how varying C from 1 μF to 10 μF in 1 μF steps affects the rise and fall times of the output waveform. V1 is a 0 to 15 volt square wave. You must choose the appropriate parameters for the pulse voltage source so that the output has time to fully charge and discharge.

Problem 6-79: For the circuit below, run a Parameter Sweep in conjunction with a Transient Analysis to see how varying R from 1 kΩ to 10 kΩ in 1 kΩ steps affects the rise and fall times of the output waveform. V1 is a 0 to 15 volt square wave. You must choose the appropriate parameters for the pulse voltage source so that the output has time to fully charge and discharge.

Problem 6-80: For the circuit of Problem 6-79, run a Parameter Sweep in conjunction with a Transient Analysis to see how varying R from 100 Ω to 1 kΩ in 100 Ω steps affects the rise and fall times of the output waveform. V1 is a 0 to 15 volt square wave. You must choose the appropriate parameters for the pulse voltage source so that the output has time to fully charge and discharge.

Problem 6-81: For the circuit of Problem 6-79, run a Parameter Sweep in conjunction with a Transient Analysis to see how varying C from 1 μF to 10 μF in 1 μF steps affects the rise and fall times of the output waveform. V1 is a 0 to 15 volt square wave. You must choose the appropriate parameters for the pulse voltage source so that the output has time to fully charge and discharge.

Problem 6-82: In the linear regulator example of section 6.L, we ran a Temperature Sweep to see how the temperature dependence of the semiconductor devices affected the output voltage of the regulator. In this problem we will add the temperature dependence of R1 and R2 to the simulation. We can add temperature dependence to a virtual resistor. A virtual resistor is one you place in your circuit and then change its resistance value. A non-virtual resistor is one of those you pick from the list of available resistors such as 1.0, 1.5, 1.8, 2.2, 2.7, and so on. R1 and R2 must be virtual resistors. If you double-click on the graphic of the resistor, you will see that there are two parameters you can set called TC1 and TC2. These parameters are the temperature coefficients of the individual resistors. Resistance as a function of temperature is given as:

$$\text{Resistance} = \mathbf{R}\left[1 + \mathbf{TC1}(T - Tnom) + \mathbf{TC2}(T - Tnom)^2\right]$$

R is the value of the resistor specified on the schematic. TC1 is the linear temperature coefficient and TC2 is the quadratic temperature coefficient. If you do not specify values for TC1 and TC2, their values are set to zero, which means no temperature dependence. Tnom is the nominal temperature of the simulation and has a default value of 27°C.

A typical resistor may have a linear temperature dependence of 100 parts per million (TC1 = 100e-6, TC2 = 0). Add this temperature dependence to R1 and R2 and simulate the circuit with the same analyses used in section 6.L.

PART 7
Digital Simulations

Multisim can simulate pure analog, mixed analog/digital, and pure digital circuits, and also has VHDL capabilities. This part describes how to run pure digital and mixed analog/digital simulations. The circuits given are fairly simple, but the examples can be applied to larger systems. We can either use the Transient Analysis simulation or run a Virtual Lab simulation.

For this section, we will assume that you have studied Parts 1 and 2, which show how to draw circuits and use the Graphical Postprocessor. We will also assume that you have done a number of the exercises in the other sections so that you are familiar with the use of the oscilloscope and know the location of most of the parts we will be using in this section.

7.A. Digital Indicators, Signal Generators, and Instruments

To demonstrate the different instruments and indicators available for use in digital simulations, we will simulate the circuit below:

This is a fairly large circuit, so we will build it slowly and test it piece by piece. It contains the Word Generator, Logic Analyzer, and Oscilloscope instruments, some indicator lights, a buzzer, a logic gate, and some LED bar graphs. These are the items you will use to simulate and observe the behavior of digital circuits.

7.A.1. Word Generator and the Logic Analyzer

We will start by creating the portion of the circuit shown below:

The Word Generator and the Logic Analyzer are located in the Instruments toolbar. To place the Word Generator, click the *LEFT* mouse button on the **Word Generator** button as shown below. The instrument will become attached to the mouse pointer:

To place the part, use the mouse to move the part to the desired location and click the *LEFT* mouse button. Place the part in a convenient location:

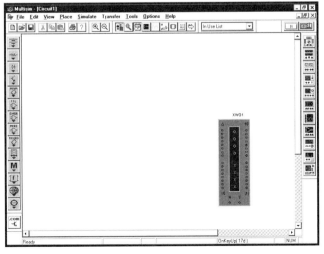

To place the Logic Analyzer, click the **LEFT** mouse button on the Logic Analyzer button as shown below. The instrument will become attached to the mouse pointer:

Place the part in a convenient location:

We will now take a closer look at the Logic Analyzer and the Word Generator graphics. The Logic Analyzer is shown below.

The Logic Analyzer has sixteen inputs labeled **1** through **F** (although they really should be 0 through F). Remember that when dealing with logic circuits, F is a hexadecimal number and represents the number 15 in base 10, or the number 1111 in binary. Either way, the Logic Analyzer has sixteen inputs that can be displayed simultaneously. The labels **1** and **F** tell us which are the first and last inputs. The little circles are wiring terminals. We can connect these circles to locations in the circuit and observe the waveforms.

An enlargement of the Word Generator is:

This instrument has 32 outputs, labeled 0 through 31. Each output generates a logic signal that can be used as an input to a logic circuit. The output of each channel is independent of the other channels.

For our example, we will be using the first four outputs of the Word Generator and the first four inputs of the Logic Analyzer. To make wiring the two instruments together easier, we will need to flip the Logic Analyzer horizontally so that the input terminals face the Word Generator. To flip the Logic Analyzer, click the **RIGHT** mouse button on the Logic Analyzer graphic. A menu will appear:

Select **Flip Horizontal**:

The input terminals for the Logic Analyzer are now on the right side of the instrument.

We can now easily wire the Logic Analyzer and the Word Generator together. If you are unfamiliar with wiring components together, see section 1.D. You may need to shift one of the instruments up or down so that the pins line up nicely:

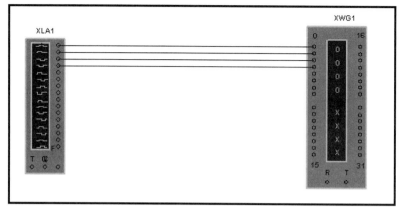

The next thing we want to add is the decoded 7-segment display. This part is located in the **Indicators** toolbar. Click the *LEFT* mouse button on the Indicators button as shown below:

Click the *LEFT* mouse button on the **HEX_DISPLAY** button to obtain a list of the available 7-segment displays:

The **DCD_HEX** part is a decoded 7-segment display. This means that you can give it four binary inputs from 0000 to 1111, and it will display an alphanumeric character of 0 through F. No other logic is needed to convert between the binary inputs and the LED segments because the conversion logic is built in to the part. The other two parts are basic 7-segment displays that have inputs that directly control a single segment. You need to add the conversion logic to control each segment. If you are studying 7-segment displays and their driver circuits, you will want to use either the **SEVEN_SEG_COM_K** or the

SEVEN_SEG_DISPLAY because you have to add all of the required circuitry to get these parts to function. If you want a quick display of 4 binary inputs, use the **DCD_HEX** part because all you need to do is place it in your circuit and connect the four inputs.

Since the part DCD_HEX is already selected, all we have to do is click on the **OK** button to place the part in our circuit. Place the part and then wire it as shown:

The last thing we must do before running a simulation is to program the Word Generator. Double-click the *LEFT* mouse button on the graphic of the Word Generator:

We would like the Word Generator to count from 0 to 15 (Fh) and then back down to zero. This is a total of 31 values (1Eh) that will use 31 memory locations in the Word Generator. We will set the **Initial** address to **0000** and the **Final** address to **001E**:

We have told the Word Generator that there will be 31 memory addresses. We must now specify the contents of each of those memory locations. We can generate the increasing count automatically. Click on the **Pattern** button:

Select the **Up Counter** option and click on the **Accept** button:

If you look at the output data, you will notice that 31 lines are listed and numbers increase from **0** to **1E** (hexadecimal), which is 0 to 31 in base 10:

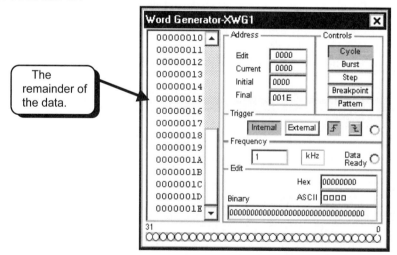

We see that we can easily create data that counts up. To create the portion of the data that counts down, we have to enter it manually. Locate address location 10h and select it. To select it, click on the data:

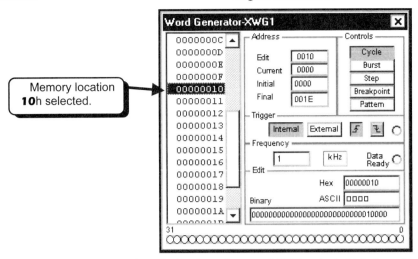

We can enter the new value as a hexadecimal number, an ACSII code (which doesn't make much sense in this example), or as a binary number. We will enter it as a hexadecimal number. Click the **LEFT** mouse button as shown below to place the cursor in the indicated field:

With the cursor in the **Hex** field, we can now change the value of the field. Use the **BACKSPACE** key to erase the value and then enter the value **E** into the field:

When you change the value in the **Hex** field, the data in the memory location changes as well.

Next, we need to change the value of memory location 11h to Dh or 1101 binary. Locate address location 11h and select it. To select it, click on the data:

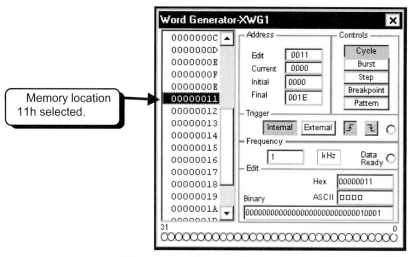

For this memory location, we will enter the value as a binary number. Click the **LEFT** mouse button as shown below to place the cursor in the indicated field:

With the cursor in the **Binary** field, we can now change the value of the field. Use the **BACKSPACE** key to erase the values that are not zero and then enter the value **1101** into the field:

When you change the value in the **Binary** field, the data in the memory location changes as well. Note that the values are displayed in Hexadecimal in the memory location to save space.

We must now change the remaining memory locations manually. Change the data in your Word Generator so that it matches the data below:

The data goes from 0 to F and then back down to 0.

Now that we have programmed the Word Generator, we need to tell it how to run. **Cycle** mode means that the Word Generator starts at the first address and outputs the contents of each memory location sequentially. After the Word Generator reaches the last memory location, it goes back to the beginning and starts over. In **Cycle** mode, the Word Generator will cycle through the data over and over until you stop the simulation. In **Burst** mode, the Word Generator will output the contents of each memory location once and then stop. In **Step** mode, the Word Generator will only output the contents of one memory location at a time. Each time you click on the **Step** button, the Word Generator will go to the next memory location and output its contents. It will remain at that memory location until you click on the **Step** button again. This mode of operation is good for debugging circuits. When programming the Word Generator, if you select a memory location and then click on the **Breakpoint** button, that memory location will be specified as a breakpoint. When you are in Cycle mode or Burst mode, the Word Generator will stop and hold the output value at a breakpoint. The Word Generator will not resume stepping through the memory locations until you click on the **Cycle**, **Burst**, or **Step** buttons. To remove a breakpoint, select the memory location with a breakpoint and then click on the **Breakpoint** button again.

We are now ready to start the simulation. Make sure that the Word Generator window is not covering the alphanumeric display and click on the **Cycle** button:

Clicking on the **Cycle** button starts the simulation. You should see the alphanumeric display go from 0 to F and then back down to zero. You will also see the current memory address that the Word Generator is outputting:

To stop the simulation we need to click on the **Run/Deactivate** button as shown:

You can observe the behavior of the Word Generator by using the **Burst** and **Step** buttons if you wish. Remember that with the **Step** button, you must click on the button each time you want the Word Generator to go to the next memory location.

The last thing we will do is observe the waveforms using the Logic Analyzer. Double-click the *LEFT* mouse button on the Logic Analyzer to open its window:

The window displays a few incorrect waveforms because the Logic Analyzer is not set up correctly. The only thing we need to do is change the clock section. Click the *LEFT* mouse button on the **Set** button to adjust the sampling frequency:

The clock rate is the sample rate of this instrument. Once every clock period, the Logic Analyzer measures the input channels and displays those values on the screen. The displayed values will remain unchanged until the next sample is taken, even if the inputs change. Thus, we need to make the sample frequency much higher than the frequency of the waveform we are measuring. Our Word Generator was set at 1 kHz, so we need to make the **Clock Rate** of the Logic Analyzer much higher than 1 kHz. Choose a frequency that is a factor of 2^n higher than 1 kHz. The reason for the 2^n will become apparent later. I will choose 16 kHz for the **Clock Rate**:

Also make sure that the **Clock Source** is specified as **Internal**. We are not providing a clock for the Logic Analyzer, so it will need to generate its own clock signal. Click on the **Accept** button when done.

The next thing we need to do is to set up the display of the Logic Analyzer. The display will be broken up into divisions, and we need to specify the number of Logic Analyzer clock pulses per division (**Clocks/Div**). The number of clocks per division can be set to 1, 2, 4, 8, 16, and so on, up to 128. This is why I chose the clock rate to be a factor of 2^n times as fast as Word Generator frequency. I will choose the **Clocks/Div** to be 16. Since the Logic Analyzer clock frequency is 16 kHz, this will make the time per division equal to 1 ms, which happens to be the period of our Word Generator frequency:

We can now run the simulation. Click on the **Cycle** button of the Word Generator to start the simulation:

The simulation will run and the data will be displayed on the Logic Analyzer screen:

If you stop the simulation by clicking on the **Run/Deactivate** button , you can use the cursors to measure values on the logic Analyzer screen. Cursor one is red and cursor two is blue. To move a cursor, drag the triangle at the top of the cursor. The procedure for moving the cursors is the same as it was for the oscilloscope in section 6.A.4. Shown below is a screen capture with the cursors:

The cursors show the time coordinates of each cursor and the time difference between the cursors. The cursors also give the value of each input at the cursors, although this is not as obvious. If you treat input 0 as the least significant bit, and channel 15 as the most significant bit, the inputs to the Logic Analyzer can be treated as a 16-bit binary number. The value of this number is displayed in hexadecimal in the cursor information:

If we look at cursor 1, we see that channels 1 through 3 are high, and channel 4 is low. The other channels are not used and are treated as a logic zero. Thus, the inputs at cursor 1 are 0111, which corresponds to 7h. If we look at cursor 2, we see that channels 1 and 2 are high, and channels 3 and 4 are low. Thus, the inputs at cursor 2 are 0011, which corresponds to 3h. This information is displayed as part of the cursor display:

We are now done with these instruments for the moment, so we need to close their windows. Click on the buttons as shown to close the windows:

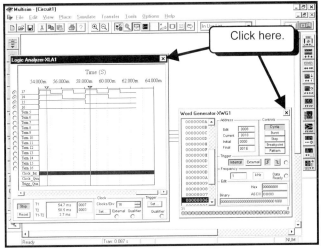

We will continue with this circuit in the next section.

7.A.2. Digital Probe and Undecoded Bar Graph Lights

We will next look at two indicators that can be used to view the status of a single bit. The digital Probe part can be directly connected to a wire, and it will indicate if the voltage on the wire is high or low. An undecoded bar graph is a package of ten LED indicator lights. You must drive these LED's as you would any LED. That is, you must add a series resistor to limit the current, and you must also provide a driver circuit if the output you are trying to measure cannot supply the current needed to drive an LED.

The digital Probe is located in the indicators toolbar. Click the **LEFT** mouse button on the Indicators button as shown:

Click on the **PROBE** button to select the part. The **PROBE** graphic will become attached to your mouse pointer and move with the mouse:

Place four **PROBE**s in your circuit as shown. To rotate a part, type **CTRL-R** after you have placed the part on your page:

Notice that the text **2.5 V** is displayed on the graphic for the Probe:

2.5 V is the threshold for the Probe. If the Probe measures a voltage less than the threshold, the Probe will be off. If the Probe measures a voltage greater than the threshold, the Probe will illuminate. The threshold is displayed on the schematic so that you can easily change it if you wish. A threshold of 2.5 V works well for our example, so we will not change it. A Probe draws no current, so you can connect it to your circuit and not worry that it will affect the operation of your circuit. Wire up the Probes as shown:

The Probes will now display the status of each output of the Word Generator. The rightmost Probe (**X4** in my circuit) is the least significant bit and the leftmost Probe (**X1**) is the most significant bit.

Next, we will add the undecoded LED bar graph to our circuit. This part is also located in the **Indicators** menu. From the **Indicators** menu, select the **BARGRAPH** button:

The dialog box displays all of the available bar graph components. We want an undecoded bar graph, which is called **UNDCD_BARGRAPH**. Select this part and then click on the **OK** button. The part will become attached to the mouse pointer:

426 Digital Simulations EWB Multisim

This bar graph is just ten individual LEDs mounted in a single package and arranged into a bar graph display. To use this bar graph we must know which terminals are the anodes and which are the cathodes of the LEDs. Remember that current for a diode flows into the anode terminal and out of the cathode terminal. It is not obvious from the symbol, but when you initially place the **UNDCD_BARGRAPH** part in your circuit, the anodes are the left-side terminals and the cathodes are the right-side terminals. Thus, current can flow from left to right in the drawing above. Note that if you rotate the part, the pin designations I just mentioned do not apply. Thus, when you first place the part, you know which terminal is which. When you rotate the part, you must keep track of the pins.

Place the bar graph in a convenient location and wire the remainder of the circuit as shown:

We are not using any driver circuits for the LEDs because the Word Generator can source enough current to illuminate the LEDs. We can now run the simulation. Click on the **Cycle** button in the Word Generator to start the simulation. The bar graph and the Probes should display binary codes that agree with the alphanumeric display. A few screen captures are shown below:

7.A.3. Mixed Signal Indicators

We will now add a buzzer to the circuit and test the other types of bar graph displays. We will also look at two of the waveforms using the oscilloscope. Finish the remainder of the circuit as shown below. This section contains a number of analog parts that have been covered in Parts 3 through 6 of this book.

The buzzer is located in the **Indicators** toolbar, and the two other bar graph displays are placed the same way as the bar graph of the previous section. The 4-input AND gate is located in the CMOS toolbar. To obtain this toolbar, click on the CMOS toolbar button as shown below:

The **CMOS** button bar shows all of the CMOS logic families available. The part I chose was from the 4V family, so click on the **CMOS 4V** button as shown:

The dialog box lists all of the four-volt CMOS parts available to us. Locate the 74HC21N and select it:

Click on the **OK** button to place the part.

To place the AC voltage source, click the ***LEFT*** mouse button on the Sources toolbar button and then click on the **AC_VOLTAGE_SOURCE** button :

EWB Multisim   Digital Simulations   429

To place the 2N2222A transistor, click on the Transistors button:

The 2N2222A is an NPN transistor, so click on the NPN Transistor button:

This dialog box lists all of the NPN transistors available to us. The one selected happens to be the one we want to use. To place the part in your circuit, just click on the **OK** button.

You should now be able to place all of the parts in the circuit and wire it up. A few of the components require some discussion. We must modify the settings of the buzzer. Double-click the ***LEFT*** mouse button on the buzzer graphic to edit its properties:

The frequency of the buzzer is 200 Hz, it requires 9 volts to operate, and it draws 50 mA of current. Our driver circuit was designed to drive a 4-volt buzzer, so change the voltage to 4 volts as shown:

You can change the frequency and current of the buzzer if you wish. However, choosing too large a current for the buzzer may cause problems with the 2N2222A transistor. Click on the **OK** button to accept your modifications. The 4-input AND gate detects when the Word Generator emits the code 1111. When this occurs, the buzzer should make a noise.

The **LVL_BARGRAPH** and the **DCD_BARGRAPH** are actually analog indicators. These indicators take a single analog input and then display the value as a level on the bar graph. These parts contain the logic circuits to convert the analog signal to 10 digital signals, and also contain the drive circuits so that those ten logic signals can be displayed on the LED bar graph. To use them, all you have to do is place them in your circuit and then connect the two inputs to ground and the signal you want to measure. For our circuit, these bar graphs are connected to a 15-volt sinusoidal voltage source, so the display should go up and down with the sinusoidal voltage.

When you have completed wiring the circuit, you can start the simulation. Click on the **Cycle** button in the Word Generator to start the simulation. The buzzer should sound every time the Word Generator hits the code 1111. The two decoded bar graphs should rise up and down as they follow the sinusoidal voltage source. A few screen captures are shown below:

The buzzer should sound when you see the following displayed on your screen:

The last thing we will look at are the waveforms displayed by the oscilloscope. See section 6.A for instructions on displaying waveforms with the oscilloscope. The waveforms are shown below:

The purpose of showing these waveforms is twofold. The first point is that we can use the oscilloscope to measure waveforms in a digital circuit. This is useful if we have a digital circuit that is connected to an analog waveform somewhere in our circuit. The second point we want to emphasize is that the AC voltage source is not synchronized to the Word Generator. If we blow up the time axis and use the cursors, we see that the edges of the Word Generator do not match up with the zero crossings of the AC voltage source:

If we want to synchronize the Word Generator to the zero crossings of the AC voltage source, we need to add some circuitry.

7.B. Mixed Analog and Digital Simulations

The first circuit we will look at is an OPAMP circuit that drives the clock of a J-K flip-flop. Wire the circuit shown below:

The circuit is drawn as though we were going to wire the circuit in the lab. No special circuits are required between the analog circuitry and the digital logic gates. Note that the J and K inputs of the flip-flop are held high so that the flip-flop toggles at each negative clock edge.

The sinusoidal voltage waveform produces a 1 kHz sine wave with voltages between –1 and 1 volts. The **741** OPAMP circuit has a gain of 11 and produces a ±11 volt sine wave of 1 kHz at node **Vo1**. This waveform goes into a Zener clipping circuit that limits the voltage to approximately +5.6 and –0.7 volts at node **Vclamp**. This voltage is TTL compatible and can be connected to the Schmitt Trigger (7414) input. The output of the Schmitt Trigger (**Vo2**) should be a 0- to 5-volt square wave at 1 kHz. The J-K flip-flop is wired as a divide-by-two counter, so Q and $\overline{Q}$ should be 0- to 5-volt square waves at 500 Hz. They should also be 180 degrees out of phase.

We will now run the simulation and use the oscilloscope to view the waveforms. Place the Oscilloscope instrument in your circuit and connect it as shown:

We will first test the input and output of the 741 OPAMP circuit. This circuit has a gain of 11, so the output should be in phase with the input, but 11 times larger. Click the **LEFT** mouse button on the Run button to start the simulation. Double-click the **LEFT** mouse button on the oscilloscope to open the oscilloscope window. Adjust the settings of the oscilloscope as shown below. If you are unfamiliar with using the oscilloscope, see section 6.A for detailed instructions:

EWB Multisim   Digital Simulations   433

We see that the input signal (**Channel A**) is a 1-volt peak signal, and the output (**Channel B**) looks to be about an 11-volt peak signal. We could use the cursors to measure the peak of the output waveform, but we will just eyeball it. Thus, it appears that our OPAMP circuit is working properly.

Next, we will change **Channel A** of the oscilloscope to measure the voltage across our Zener clamp (**Vclamp**):

Restart the simulation and display the waveforms:

Both channels are set to 5 volts per division. We see that the Zener clipping circuit limits the voltage to a little over 5 volts and about –0.7 volts. This is the expected operation of this clipping circuit.

Next, we will change **Channel B** to measure the output of the 7414 Schmitt trigger (**Vo2**):

The waveforms at this point are:

Note that we have changed the **Y position** fields in the Oscilloscope settings to separate the two traces. We see that the Schmitt trigger takes an input pulse with slow rise and fall times and converts it into a pulse with very sharp edges. This is the purpose of using the Schmitt trigger. The output of the Schmitt trigger is now a suitable signal that can be used for a logic input to a digital circuit, including the clock input of a flip-flop.

The last thing we will do is observe the output of the JK flip-flop. The flip-flop is wired as a divide-by-two counter, so the Q output of the flip-flop should be a square wave of half the frequency of the clock input. Change the oscilloscope connections as shown and then observe the waveforms:

In order to display the waveforms, we must set the trigger **Level** of the oscilloscope to a voltage greater than 0 and less than about 4 volts. If we do not make this change, the oscilloscope will never trigger because neither of the measured waveforms ever goes below zero, which is the default trigger level.

For a second example, we will simulate the circuit below:

In this circuit, we are using the Q output of the 7476 as the clock input of the two 74160 counters. The 74160s are decade counters, which means that they count from 0 to 9. The two counters are cascaded so that they count from 0 to 99. If you click on the Run button, you should observe that the counters count from 0 to 99.

We have also connected the Logic Analyzer so that we can view the waveforms. Some sample waveforms are shown below:

Note that the Logic Analyzer clock frequency is set to 16 kHz (click on the **Set** button to change) and the **Clocks/Div** is set to 64.

EXERCISE 7-1: Design a 60 Hz sync circuit. The input to the circuit is a 12 V amplitude, 60 Hz sine wave. The output of the circuit should be a 1 ms 5 V pulse that occurs when the sine wave crosses zero with a positive slope.

SOLUTION: Multisim does not have a 74xx123 one-shot. However, in the **Mixed** toolbar there is a monostable one-shot that appears to be nearly equivalent to a 74xx123 one-shot. We will use this monostable one-shot instead of a 74xx123.

EWB Multisim — Digital Simulations — 437

The screen capture above shows that the pulse width is not quite 1 ms. We can increase the pulse width by making **R3** or **C1** larger.

7.C. Pure Digital Simulations

Multisim can be used as a logic simulator as well as a mixed analog/digital simulator. We will now simulate a switch-tail counter. Wire the circuit as shown:

We note that there are no power and ground connections in the circuit. Multisim automatically connects the power and ground for us. Multisim has two modes for simulating circuits with digital components, ideal and real. Ideal mode is faster, but it does not take into account power supply variations, device tolerances, and rise and fall times, and it does not require you to specify power and ground connections. Note that ideal mode does include gate propagation delay times. Real mode produces a more accurate simulation in terms of switching speed and signal levels, but it requires you to use the digital ground (⏚) and specify a supply ($\overset{VCC\ 5V}{\top}$). The default mode of operation is ideal mode. To change to real mode, you must specifically select the option, which we will not show here. We will discuss the difference between these two modes in section 7.E.

If you are familiar with a switch-tail counter, you can click on the Run button and observe the operation with the Probes. We cannot demonstrate the operation of the circuit in screen captures, so we will add the Logic Analyzer and display the waveforms you should see with this circuit:

With the Logic Analyzer, the pattern produced by this counter is quite distinctive and easy to recognize when it is correct:

EXERCISE 7-2: Design and simulate a 4-bit decade counter.
SOLUTION: Wire the circuit as shown. A 250 Hz clock is used.

When you click on the Run button , the alphanumeric display will cycle through the digits 0 through 9. Note that this realization is not a minimal design and there are several different ways of realizing a counter. Your design may be different.

EXERCISE 7-3: Construct and simulate a circuit that uses a 555 astable multivibrator as the clock and a decade counter to drive a decoder.

SOLUTION: Wire the circuit as shown.

Because of the large number of signals, we will display the results with the Logic Analyzer.

7.D. Startup Clear Circuit

The initial state of a flip-flop in a Multisim simulation is not always known. We will now present a technique that can be used to initialize flip-flops in both simulations and in circuits that you build. This technique is relevant because real flip-flops do not automatically set themselves to a specific state at startup. The circuit designer must add a circuit that presets or clears all flip-flops so that the circuit starts in a known state. The circuit below can be used to preset or clear a flip-flop at power-up as long as the preset or clear input is active low. Here we show how to clear two flip-flops at the start of the simulation.

Resistor **R1** and capacitor **C1** form the initialization circuit. The initial condition of the capacitor is specified as 0 volts. When the simulation runs, the capacitor starts at 0 volts (logic zero) and charges to 5 volts. While the capacitor is close to 0 volts, it provides a logic zero to the clear input of the flip-flop. For the remainder of the simulation the capacitor is charged to 5 volts, providing a logic 1 to the flip-flop clear input, which has no effect.

To specify an initial condition for a virtual simulation, double-click on the wire between **R1** and **C1**:

Change the **Node name** to `Clear`, enable the option **Use IC for Transient Analysis**, and make sure that the initial condition is zero:

Click on the **OK** button to accept the changes. When you run the simulation, no matter what the initial states of the flip-flops are, they will be cleared. For this circuit, the flip-flops are initially cleared, then they start counting since they are wired as a two-bit counter:

If you start and stop the simulation several times, you will see that the flip-flops always start at zero. We can look at the capacitor voltage with the oscilloscope:

We see that at the very beginning of the simulation, the capacitor voltage is close to zero volts. It is at a low enough voltage for a long enough period of time that the flip-flops respond to the low voltage on the clear input and clear the flip-flops. For the remainder of the simulation, the capacitor voltage is a logic high, and this has no effect on the flip-flops because the clear inputs are active low.

The trick to this circuit is that the capacitor and resistor have to be large enough that the capacitor takes a long enough time to charge up. The larger **R1** and **C1** are, the longer the capacitor spends at a low voltage, and the longer the flip-flops have to react to the low voltage on their clear inputs. If **R1** and **C1** are too small, the RC transient will be too fast and not all of the flip-flops will be fast enough to react to the low input. A better procedure is to use this RC startup circuit to trigger a one-shot. The one-shot then emits a pulse of fixed and known duration that clears or presets all flip-flops.

This startup clear circuit only works at power-up because before power is turned on, the capacitor discharges to zero. When the circuit is running, there is no way to discharge the capacitor unless you add more circuitry. Thus, with the shown implementation, the only time this circuit has an effect is at power-up or at the start of the simulation.

EXERCISE 7-4: Design and simulate a 4-bit ring counter. The counter should be initialized to 1000.

SOLUTION: Wire the circuit as shown. A 1 kHz clock is used. Use the capacitor startup circuit to preset the first flip-flop to 1 and the remaining flip-flops to 0. Note that the initial condition of the capacitor is set to zero.

When you run the simulation, you will see a single illuminated Probe circulate around the counter from left to right. We can also view the states of the flip-flops with the Logic Analyzer:

We see that only one flip-flop at a time has a logic 1 output, and this logic 1 is passed from flip-flop to flip-flop.

7.E. Real and Ideal Mode Digital Simulations and Gate Delays

Multisim includes gate delays in its digital simulations. To illustrate gate delays, we will create a two-phase, non-overlapping clock using gate delays. We will show this simulation using Multisim's ideal and real digital simulation modes of operation. Both modes include gate delays. In ideal mode, gate outputs produce square edges. Only the delay time between input and output is modeled. In real mode, gate delays as well as rise and fall times are modeled, and gate outputs are not square. Real mode requires more simulation time to model the gate properties, and thus simulations take longer to complete.

7.E.1. Ideal Mode Digital Simulations

The default mode for digital simulations is the ideal mode. If we proceed as we did in previous simulations, we will be using ideal mode, so we do not need to change anything. In ideal mode you can use either ground for your drawings (⏚ or ⏚). Wire the two-phase clock circuit below:

The output of U1D should be the same as the input to inverter U1A. The only difference between the signals is the delay time through the four inverters. Since gate delays are on the order of nanoseconds, we will set the clock frequency to 10 MHz. We will first observe the output of the circuit, Phase 1, and Phase 2. Two oscilloscope windows displaying the same waveform at different time scales are shown:

We see that the circuit does produce two out-of-phase clock pulses from a single 10 MHz clock input. The circuit functions because of the gate delays of the four inverters. We can measure the delays caused by the inverters with the oscilloscope connected as shown:

We can use the oscilloscope cursors to measure the delay through the four gates. See section 6.A.4 for using the oscilloscope cursors. Two screen captures with different time scales are shown:

The delay for four inverters is about 75 ns, so the delay per inverter gate is approximately 19 ns.

7.E.2. Real Mode Digital Simulations

The one thing we notice about all of the previous simulations is that the edges of the digital waveforms are square. The logic and delay times of the devices are modeled correctly, but in real devices the edges of output waveforms are not square. We will simulate the same circuit as in the previous section, but we must make a few modifications so that it will work in real mode. Before we change the circuit, we will look at the analog ground symbol. Double-click the *LEFT* mouse button on the ground wire, and you will see that using the ⏚ part names the ground node zero (0).

The node name is shown in green and may be hard to see in the screen capture. A ground name of 0 is the standard for SPICE simulations. For real mode simulations, we have to use the digital ground. Modify the circuit as shown below:

We see that we have added a Vcc part that is not connected to anything. In analog circuits, this part was used to connect other circuit elements to a power supply of the specified voltage. Here, it is not connected to anything, but it does set a node called **VCC** to **5** volts. All digital circuits are connected to this node even though we do not specifically wire the gates to the node. Thus, we only need to place the part in our circuit. Since the power pin of logic gates is connected to node VCC, this part is connected to the gates even though we do not draw a wire.

We have also replaced the analog ground by the digital ground symbol (▽). If we now click on the ground wire, it will have a different label:

We see that the ground node is now named **VGND** rather than 0. In real mode, the default power and ground nodes are **VCC** and **VGND**. It is possible to change the power and ground names, but we will not do this here.

Next, double-click on a digital part such as an inverter:

We see that the power and ground nodes are listed as **VCC** and **VGND**. If you want to change them, you can use this dialog box. We do not need to change them, so click on the **Cancel** button.

Before we run the simulation, we need to switch to real mode. Select **Simulate** and then **Digital Simulation Settings** from the menus:

This dialog box allows us to select between real and ideal modes. We see that the default mode for digital simulations is ideal mode. To switch to real mode simulations we need to select the option here. Select the **Real** option and then click on the **OK** button:

We are now ready to run the simulation. We will first look at the delay through the four inverters:

We see that the output of the digital gate has a rise time, and it also has overshoot. The total delay through the four inverters is measured as approximately 76 ns, nearly the same as in the ideal mode simulations. The fall time of the inverter output is very fast and does not have an RC decay or undershoot.

Next, we will look at the output of the circuit:

The waveforms are:

We see that the circuit still develops a two-phase clock, but the pulses are not nearly as clean as in the ideal simulation. The pulses shown above are what you would expect to measure if you built this circuit in the lab. The width of the pulses is about the same as we saw with the ideal mode simulations.

7.F. Problems

Problem 7-1: Create a 1 kHz square wave with a 50% duty cycle using a 555 timer and a JK flip-flop.

Problem 7-2: Create a 10 kHz square wave with a 50% duty cycle using a 555 timer and a JK flip-flop.

Problem 7-3: Using a 74xx160 counter, create a counter that counts from 0 to 5. Use the alphanumeric display to display and verify the operation of your circuit. Use the startup clear circuit covered in section 7.D to initially clear the counter.

Problem 7-4: Using a 74xx160 counter, create a counter that counts from 3 to 9. Use four Probes to display and verify the operation of your circuit.

Problem 7-5: Create a 10-bit ring counter and display the outputs on a bar graph. Use the startup clear circuit covered in section 7.D to initialize the counter.

Problem 7-6: Create a binary counter that counts from 0 to 255. Connect the 8 outputs of this counter to a DAC and add the necessary extra components so that the analog output is a voltage from 0 to 5 volts. Show that if you let the counter continuously cycle from 0 to 255, the analog output will be a ramp from 0 to 5 volts at a frequency that is equal to the clock frequency divided by 256.

Problem 7-7: Create a binary up/down counter that counts from 0 to 255 and then down to 0. Connect the 8 outputs of this counter to a DAC and add the necessary extra components so that the analog output is a voltage from 0 to 5 volts. Show that if you let the counter continuously cycle, the analog output will be a triangle wave from 0 to 5 volts at a frequency that is equal to the clock frequency divided by 510.

INDEX

#

⏚ 29, 140
A button 125
AC button 126
Analog button 248
Analog button 329
BARGRAPH button ... 425
Basic parts button 327
Bode Plotter button 231
CMOS 4V button 428
CMOS toolbar button 427
digital ground symbol ... 445
digital Probe off............ 425
digital Probe on 425
Diodes button 328
FWB (Full-Wave Bridge) button 329
Hex_Display button ... 414
Indicators button 115
key 2
Logic Analyzer button 412
maximize button 4
maximize icon 49
Multimeter AC button 215
Multimeter button 122

Multimeter current button 215
Multimeter icon 214
Multimeter voltage button 215
Multisim 2001 3
NPN_BJT button 244
OPAMP button 248
oscilloscope button 293
Piecewise_Linear_Voltage_Source button 354
Probe button 424
Pulse_Voltage_Source button 362
simulate button ..... 118, ... 125
Sources button 249
Transformers button .. 328
Transistors button 151, 243
V button 125
Vcc button 250
voltage indicator button 115
Wattmeter button 129
Word Generator button 411
Zener diode button 355
$\mu A/V^2$ 187
μV 10

@

@qq1[ic] 192, 196
@rr1[i] 179

0

0 - ground 32

1

10 to 1 probes 294
1N4001GP 181
1N4734 355

2

27°C 145
2N3904 .... 152, 193
2N3904 transistor 243
2N3906 199

3

–3 dB frequency 257, 264
3-terminal plug 294

5

5% resistor 16
555 timer 439

6

60 Hz ripple 333

7

7414 432
74160 435
74HC21N 428
7815 linear voltage regulator 329

9

90° Clockwise 18

449

A

A and B terminals 294
A button 125
absorbed power 129
AC Analysis 217
   Bode plots 235
   Decade Sweep type 256
   Decades 237
   Decibel 237, 256
   Linear 237
   Linear Sweep Type .. 220, 237
   Logarithmic 237
   Number of Points 220
   Octave 237
   Octave sweep type 237
   Output variables tab 220
   Plot during simulation button
      221
   points per decade 237, 256
   Simulate button 221
   Start frequency 220
   Stop frequency 220
   Sweep Type 220
   Vertical scale 220, 256
AC button 126
AC coupling 299, 335
AC Frequency Analysis 218
AC Mode 213
AC portion of signal 333
AC power 265
AC Source
   Analysis Setup tab 236
AC Sweep 217
   Vertical scale 237
AC voltage source 6, 7
Add device/model parameter
   button 156, 175
Add Trace button . 48, 50, 55, 60
adding a trace 48
admittance 267
amplifier gain 243, 255
amplifier voltage swing 368
amplitude 292, 369
analog oscilloscope 292
Analog toolbar 329
Analysis Results 170
Analysis Setup tab 218, 236
Analysis to sweep 262
AND gate 427
angle of load 265
Apply button 90
asterisk 142
audit trail 152
auto triggering 305
auto-hide 2
Autorange 93
Autowire 23
average power 63
axis label 87

B

bandwidth 259
bar graph 423
base current 155
Basic button 12
binary 417
bins 5
bipolar junction transistor 6, 151
BJT 6
   characteristic curves 191
   DC current gain 194
   emitter current 194
   HFE 194
   I-V characteristic 191
   operating point 151
   PNP 199
   VCE 200
BJT inverter 395
Bode plots 230, 235
   examples 239, 241
   phase 230
Bode Plotter
   Grapher 233
   instrument 230
   Magnitude button 232
   Vertical axis 245
   x-axis range 232
   y-axis range 232
bottom axis 87
Bottom Axis tab 89, 197
branch current 170
buttons with a gray background
   13
buttons with a green
   background 13
buzzer 427, 429

C

capacitor 6, 13
   initial condition 381
change name 11
Change Part Model button . 142, 188
Channel A minus Channel B 318
characteristic curves
   BJT 191
   jFET 193
checkmark 134
Circuit tab 134
clipping circuit 361, 367
clock
   pulses 444
   two-phase 443
Clock Rate 421
Clock Source 421
Clocks/Div 421
clockwise 18
Close button 49, 51
close toolbar 12, 22
closed-loop gain 259
CMOS inverter 190
CMOS logic gates 7
collector current .. 155, 158, 192
collector-emitter voltage 192
common-base amplifier 246
common-emitter amplifier .. 243
Comparators 6
Component Bin 5
Component Family .. 19, 20, 330
Component Name List 248
connecting
   components 23
   to a corner 29
   wires 27
connection 27, 31
Copy Variable To Trace 48
corner 26
correcting mistakes 22
correcting wiring mistakes 33
counter 432, 435
coupling 299
creating waveforms 356
crosshairs 24, 27, 30
CRT screen 302
CTRL-ESC 2
CTRL-G 49, 51, 70, 111
CTRL-R 18
CTRL-S 35
CTRL-W 19
current 314
current convention 117
   Multimeter 123
current direction 117
current indicator . 115, 116, 149, 210
   current direction 117
   ideal 120
   Mode 117
   series resistance 117
current measurement 124
current monitor 171, 174, 313
current sensing resistor 314
current sensor 228

current sources 6, 7
current through resistor 175
current-controlled current
    source 139
current-controlled voltage
    source 171, 174, 217, 219,
    313
cursor information . 98, 101, 308
cursors 95, 257, 325
    moving 98, 307
    oscilloscope 306
Cursors On 97
cut button 52, 59

D

damped sine wave 386
Damping Factor 386
dashed line 24, 26
data file 356
DC coupling 299
DC current gain 194
DC Mode 213
DC node voltages 132
DC Operating Point 151, 156
DC Operating Point Analysis
    .... 115, 132, 137
DC power supply . 327, 331, 351
DC supply ripple 344
DC Sweep 167
    nested 191
DC transfer curves 185
DC voltage source 6
DCD_BARGRAPH 430
DCD_HEX 414
dead-zone clipping circuit .. 361, 367
decade counter 435, 438
Decade Sweep type 256
Decades 237
Decibel 237, 256
decoded 7-segment display . 414
decoder 439
Decrease Zoom button ... 18
default instrument settings ... 144
default setting of indicator ... 213
Delay Time .. 363, 381, 386, 390
delays 444
delete a page 52
Delete button 71
Delete Diagram button 76
delete graph 74
delete Grapher page 77

Delete Page button 78
Delete Trace button 73
dependent sources 6, 7, 136
depletion MOSFET 187
Description field 176, 261
device model parameters 157
Device Parameter 259
Device Type 176
digital ground .. 29, 33, 438, 445
digital Probe 423
digital simulations 410, 437
digital waveforms 445
diode 6
    bridge 6
    current 140
    emission coefficient 140
    Is 141
    I-V characteristic 180
    saturation current 140, 141
    Toolbar 140, 181
    Virtual 140
    Virtual button 141
    Zener 354
Diode_Virtual 140
Diodes Incorporated 327
display node names 33
display title block 35
distortion 371
distortion-free amplifier 372
divide-by-two counter 432
Divisions 93
dot at connection 27
drain current 161
Draw button 48, 51, 52
drawing a circuit 1
drawing a wire 24

E

edit
    clear menu selection 71
    Model button 141, 188
    properties 8, 9
emission coefficient 140
emitter current 194
End time 339
enhancement MOSFET 187
ESC key 24
Excel 357
exponentially damped 386
extension
    gra 110
    pgf 67
external trigger 294

F

F5 key 119, 120, 148
F8 key 8, 22
F9 key 8, 37
Fall Time 363, 390
File
    Recent Files 67
    Save 35
Filter Unselected Variables . 161
flip 413
flip component 18
Flip Horizontal 413
flip-flop 434
footprint 13
Fourier Analysis 372, 375
    Edit transient analysis 373
    Fundamental Frequency .. 372
frequency 11, 386
    components 374
    fundamental 372, 374
    response 230

G

GΩ 119
gain 243
gain bandwidth product 257
gate delays 444
GBPC2502 327
GBPC2502A 327
GBW 258
general ground symbol 29, 32
General Semiconductor 327
General tab 81, 88, 97
Generate time steps
    automatically 339
gra extension 110
Grapher 43, 49, 79
    Apply button 90
    Autorange 93
    axis label 87
    bottom axis 87
    Bottom Axis tab 89
    cursor information 98, 101
    cursors 95
    Cursors On 97
    cut button 52, 59
    delete all pages 71
    Delete button 71
    delete page 52
    Divisions 93
    Draw button 52
    Edit Clear 71

error 52
File and Save 110
General tab............. 81, 88, 97
Graph Properties 97
Grid.................................... 82
Label 86, 91
Label text field 84
left axis 87
Left Axis tab 88
Legend 80, 82
move cursors 98
numerical values 95
opening 70
page 46
Pen Size 83, 86, 91
Properties 81, 97
Range 93
Restore Graph 104
Right Axis 92
Right Axis tab 93
save and load graphs 110
second y-axis 90
tabs 52
Title 88
Toggle Cursors 96
trace color 82
Trace field 85
trace labels 82
trace width 82
Traces tab.................... 83, 91
two y-axes 90
x-axis label 87, 89
x-axis scale 104
y-axis label 87, 89
y-axis scale 104
y-axis settings 88
zoom in 105
zoom out 104
zoom rectangle 103
zooming 101, 370
gray background 13
gray resistor 13
gray resistor button 15
green background 13
Grid.. 82
ground clips 294
ground connections 7
ground lug 294
ground name 445
ground symbols 6, 29
ground terminal (G) 294
grounding.............................. 29
Group all traces on one plot 262, 393

H

half-wave rectifier 336, 351
harmonic distortion 374
harmonics 374, 375
hexadecimal........................ 412
HFE 194, 199
hover...................................... 7
hybrid-π 155
hysteresis curve 388

I

ideal mode 443
ideal mode digital simulation
 437
ideal operational amplifier .. 379
ideal parts 141
impedance measurement 269, 271
 active circuit 275
 passive circuit 271
independent voltage source 6
indicator...................... 115, 423
 AC Mode 213
 DC Mode 213
 default setting 213
 upper frequency limitation
 210
Indicators button................. 211
Indicators toolbar 115, 211
inductive circuit.................. 313
inductor current 314
inductors 6, 13
initial condition 440
 capacitor 381
Initial Value................. 363, 381
initialization circuit 440
input terminals of oscilloscope
 293
Instruments toolbar...... 214, 266
integrator 382
Internal 421
International Rectifier 327
inverter 187, 395
I-V characteristic 180, 183
 BJT 191
 jFET 193

J

jFET.............................. 6, 275
 characteristic curves 193
 I-V characteristic 193
J-K flip-flop....................... 432

K

kΩ 14
knurled knob 294
kOhm................................... 14
Kp 187
kV 10

L

Label 86, 91
Label tab................... 11, 14, 17
Label text field 84
labeling nodes 33
lagging........................ 311, 316
lambda 187
leading 311, 316
LED 6
 display 414
 indicator lights................ 423
 I-V characteristic 184
left axis 87
Left Axis tab 88
Legend........................... 80, 82
Level 303
linear
 load................................. 265
 model 255
 network........................... 230
 sweep.............................. 393
 voltage regulator 19, 329, 396
Linear 197, 237
Linear Sweep Type 220, 237
linearized model 243
List 399
list of temperatures.............. 399
List Sweep Variation Type . 261
LM124J 248, 259
LM7815CT 331
load graphs 110
Load Pages button................ 69
loading Postprocessor pages . 66
Logarithmic 197, 237
logarithmic x-axis 197
Logic Analyzer.... 412, 420, 438
 Clock Rate...................... 421
 Clock Source 421
 Clocks/Div 421
 Internal 421
logic signal 413
logic simulations 437
lower –3 dB frequency........ 246
LVL_BARGRAPH 430

M

MΩ 121
magnitude 292, 314
magnitude and phase ... 210, 265
Magnitude button in Bode
    Plotter 232
magnitude function 225
magnitude versus frequency 230
maximize button 4
maximize icon 49
maximum time between
    simulation points 342
Maximum time step 340, 369
Maximum time step settings 339
MDA2502 327
MDA25021 329
mean 63
measure
    current 124
    impedance 269, 275
        active circuit 275
        passive circuit 271
    phase 309, 311
    resistance 124
    voltage 124
measured data 356
measurement error 120
midband gain 257, 258
Miller integrator 382
Minimum number of time
    points 339
minus – oscilloscope 318
Miscellaneous Options tab .. 144
mixed analog/digital
    simulations 410, 432
Mode 117, 118
model
    MOSFET 187
    parameter 259
Modify Title Block 36
monostable one-shot 436
More>> button 152, 156, 175
MOS_3TDN_VIRTUAL 187
MOS_3TEN_VIRTUAL 187
MOSFET 6
    depletion 187
    Kp 187
    lambda 187
    load 187
    model 187, 190
    threshold 187
    VTO 187

W/L 190
move cursors 98
Multimeter 115, 122, 210
    AC measurements 214
    button 122
    current convention 123
    resistance measurement ... 270
    Set button 124
    settings 124
Multisim 2001 3
Multisim Master database 19
mV 10
MV 10

N

nΩ 120, 149
negative feedback 247
nested DC sweep 191
New Chart button 47, 223
New Graph button 47, 54, 55,
    59, 60
new Grapher page 46
New Page button 46, 59
NMOS inverter 187
nodal analysis 115
    bias point 151
    dependent sources 136
node
    0 32
    display names 33
    naming 33
node numbering 33
node numbers 132
node voltage
    analysis 115
non-linear sources 6
Normal triggering 303, 305
Norton Equivalent 146
NPN bipolar junction transistor
    152
NPN pass transistor 397
NPN transistors 244
N-type depletion MOSFET . 187
N-type enhancement MOSFET
    187
Number of Points 220
numerical values 95

O

Octave 237
Octave sweep type 237
offset voltage 380

ohmmeter 270
one-shot 436
OPAMP .. 6, 247, 248, 259, 379,
    385
    frequency response.. 259, 264
    offset voltage 380
    quad package 249
    upper frequency limit 257
open circuit voltage 147
operating point 151
Operating Point Analysis 132
Option Value 145
Options
    Modify Title Block 36
    Preferences 4, 23, 33, 35, 133
oscilloscope
    0 299
    10 to 1 probes 294
    A and B terminals 294
    AC coupling 299, 335
    auto triggering 305
    coupling 299, 333
    CRT screen 302
    cursor information 308
    cursors 306
    cursors - moving 307
    DC coupling 299
    external trigger 294
    ground clips 294
    ground input 299
    ground terminal (G) 294
    horizontal axis 292
    input terminals 293
    instrument 292, 293
    Level 303
    Normal triggering 303, 305
    probes 294
    Scale 296, 322
    single trigger 305
    subtraction 318
    time coordinate 308
    time per division 292
    Timebase 292, 295, 334
    triggering 302
    volts per division 292, 297
    window 295, 310
    Y position 299, 321
    Y/T button 296
output ripple 401
Output variables tab 44
output versus input 185

P

pΩ 120
parallel RLC 327
Parameter 176, 261
Parameter Sweep . 259, 390, 395
    AC 259
    Analysis to sweep 262
    Description 261
    frequency response 259
    gain bandwidth product ... 259
    Group all traces on one plot
        .... 262, 393
    List Sweep Variation Type
        261
    Sweep Variation Type 261
part bins 5
part properties 8, 9
parts
    ground 29
    jFET 275
parts toolbar 5
pass transistor 397
passive components 13
peak amplitude 211, 218
peak value 211
Pen Size 83, 86, 91
Period 363
PF 265
pgf extention 67
phase 210, 314, 386
    function 225
    measuring 309, 311
    plots 230
    versus frequency 230
phasor 210, 236
piece-wise-linear 354
Place
    part 8, 16
    part with menus 19
    Place Component 19
    resistor 12
    single component 5
Plot during simulation .. 44, 135, 153
PMOS enhancement device. 190
PNP bipolar junction transistor
    199
points per decade 237, 256
Postprocessor 1, 43
    add trace 48
    Add Trace button 55, 60
    Close button 51
    Delete Diagram button 76
    delete graph 74

delete page 77
Delete Page button 78
Delete Trace button 73
Draw button 48, 51
Load Pages button 69
magnitude function 225
new chart 47
New Chart button 223
new graph 47
New Graph button 54, 55, 59, 60
new page 46
New Page button 54, 59
phase function 225
Save Pages button 66
saving and loading pages ... 66
vm function 225
voltage magnitude function
    225
voltage phase function 225
vp function 225
window 51
power 63, 265
power absorbed 129
power factor 265
power factor correction 267
power meter 128, 131
power supply 331, 351
power-up 442
Preferences 5, 24
Probe 294, 423
Properties 97, 183
pulse voltage source 375, 381
Pulse Width 363, 390
Pulse_Voltage_Source 362
Pulsed Value 363, 381
pulsed voltage source .. 366, 379
PWL 354

Q

Q output 434
quad OPAMP package 249

R

Range 93
real mode 445
real mode digital simulation 437
rectangular coordinates 265
reference ID 253
regulated power supply 331
relays 6
rename a node 34
resistance measurement 124, 269, 271

resistor 6, 12, 13, 20
    current sensing 314
    name 14
    properties 14
RESISTOR button 15
resonant frequency 318
Restore Graph 104
Right Axis 92
Right Axis tab 93
RIGHT mouse button 18
ring counter 442
ripple 333, 344, 402
Ripple Rejection Ratio 335
Rise Time 363, 390
RLC circuit................... 318, 327
RMS 9
    magnitude 218
    value 211
rotate component 18
Rth 147
running time average 63

S

saturation current 140
Save as Default 5, 24
save button 35
save circuit 67
save graphs 110
Save Pages button 66
save plots 66
saving a schematic 35
saving Postprocessor pages ... 66
Scale - oscilloscope 322
Schmitt Trigger .. 385, 389, 432, 433
SCRs 6
second harmonic 374
second y-axis 90
sensing terminals 319
series resistance of current
    indicator 117
series RLC circuit 318
Set button 124
SEVEN_SEG_COM_K 414
SEVEN_SEG_DISPLAY ... 415
seven-segment display 414
sharp edges 434
SHIFT-CTRL-R 18
short circuit current 149
Show all device parameters 153
Show node names 134
Show reference ID option ... 252
Show Simulation Error
    Log/Audit Trail 154

Show values option 252
Simulate.......................... 136
   Analyses
      AC Analysis 220, 236
      DC Operating Point.... 135, 152, 156
      DC Sweep.... 167, 175, 181
      Fourier Analysis 372
      Parameter Sweep . 259, 392
      Temperature Sweep..... 398
      Transient Analysis.. 43, 68, 339, 341
   Default Instrument Settings 144
   Digital Simulation Settings 446
   Postprocess 46, 50, 53, 59
   Postprocessor 66
   Run 119
simulate button 45, 118, 125
simulation points 359
single frequency AC 210
single trigger....................... 305
sinusoidal steady state 210
sinusoidal voltage source.... 368, 386
slew rate............................. 386
small-signal gain................. 243
small-signal model 243
solid line 25
source 2.............................. 200
source follower 276
Sources 6
Sources parts bin 7, 174
SPICE AC Analysis..... 217, 235
SPICE naming convention .. 168
SPICE results graphically...... 43
SPICE simulation 44
SPICE Transient Analysis ... 292
square-wave voltage source. 361
standard 5% resistor 16
standard components 13, 17
standard resistor button 15
standard value...................... 17
start drawing wire 25
Start frequency 220
Start menu.............................. 1
startup clear 440
Stop frequency.................... 220
stop simulation 138
straight lines 356
subcircuit 161
Sweep Parameter 259
Sweep Type........................ 220

Sweep Variable Type 399
Sweep Variation Type......... 261
switches 6
switch-tail counter 437
sync circuit 436

T

temp option......................... 144
temperature................. 140, 144
      dependence 401
      linear regulator 396
Temperature Sweep..... 396, 398
      Sweep Variable Type 399
ten to one probes 294
TeraOhm 148
text file of data points......... 356
THD.................................... 374
thermal voltage................... 140
Thevenin equivalent 146
threshold............................. 425
threshold voltage 187
time average 63
time coordinate................... 308
time domain analyses 292
time per division................. 292
Timebase 292, 295, 334
Title 88
title block............................. 35
      changing 37
      display 36
      modifing 36
TMAX 340
Toggle Cursors 96, 257
Toggle Legend 80
total harmonic distortion 374
trace
      color................................. 82
      label................................. 82
      width................................ 82
Trace field 85
Traces tab 83, 91
transfer curve...... 185, 360, 361, 365, 366, 367, 388
      NMOS inverter................ 187
transformer 6, 328
Transient Analysis....... 292, 339
      End time 339
      Generate time steps automatically 339
      Maximum time step. 340, 369
      Maximum time step settings 339

      Minimum number of time points................................. 339
      TMAX 340
      TSTART.......................... 340
      TSTOP............................. 340
transistor............................. 151
      operating point 151
Transistors toolbar 151, 243
TRIACs.................................. 6
triangle wave........ 357, 361, 366
trigger Level................ 303, 435
Triggering an oscilloscope.. 302
TS_PQ4_24........................ 328
TSTART 340
TSTOP 340
TTL compatible 432
TTL logic gates 6
two y-axes 90
two-phase clock 443

U

undecoded bar graph ... 423, 425
units field 10
unity power factor 267
upper –3 dB frequency....... 246, 257, 258
Use IC for Transient Analysis 441
Use this option 145
User database 19

V

V button 125
Value tab 148
Vcc part.............................. 445
VCE............................. 192, 200
Vertical axis in Bode Plotter 245
Vertical scale....... 220, 237, 256
VGND 446
VHDL 410
View 8
      Log/Audit Trail 154
      Restore Graph 104
      Show Grapher 49, 51
      Show Simulation Error ... 154
      Zoom Out 18
virtual 141
virtual capacitor 13
virtual components............... 13
virtual resistor 13
Virtual_Resistor 13
vm function 225

Voltage Amplitude 10, 386
voltage indicator 115, 210
    Mode 118
    parallel resistance 138
    resistance 118
voltage indicator button 115
voltage indicators 115, 137
    impedance 138
voltage magnitude function . 225
voltage measurement 124
Voltage Offset 386
voltage phase function 225
voltage regulator 333, 338
Voltage RMS 9
voltage sources 7
voltage swing 368
Voltage_Regulator 330
voltage-controlled current
    source 140
voltage-controlled voltage
    source 137, 319
VOLTMETER_V 212
volts per division 292, 297
vp function 225
Vth 147
VTO 187

W

Wattmeter 129, 265, 266
waveforms versus time 292
width-to-length ratio 190
Windows Start menu 1
    auto-hide 2
wire 25
    connection 27
    corner 26
    correcting 33
wiring 32
    components 23
    options 23
Wiring tab 23
Word Generator 411
    Binary field 418
    Breakpoint button 419
    Burst mode 419
    Cycle mode 419
    Final button 415
    Hex field 417
    Initial button 415
    Pattern button 416
    programming 415
    Step button 419
    Up Counter 416
Workspace 36

X

x-axis
    label 87, 89
    linear 197
    logarithmic 197
    range in Bode Plotter 232
    scale 104

Y

Y position 299
Y/T button 296
y-axis
    label 87, 89
    range in Bode Plotter 232
    scale 104
    settings 88

Z

Zener 6
    clipping circuit 185, 361, 432,
    433
    diode 354
    voltage reference 397
zero 32
zero crossing 324, 431
zoom
    box 370
    Grapher 101, 370
    in 8, 22
    in with Grapher 105
    out 8, 18, 37
    out in Grapher 104
    rectangle 103

TK7867 .H462 2004
Herniter, Marc E.
Schematic capture with Electron
Multisim /
31516000874170

SAUK VALLEY COMMUNITY COLLEGE
LIBRARY
DIXON, IL 61021